LTL

ALLEN COUNTY PUBLIC LIBRARY
3 1833 04291 6442

P9-AFR-505

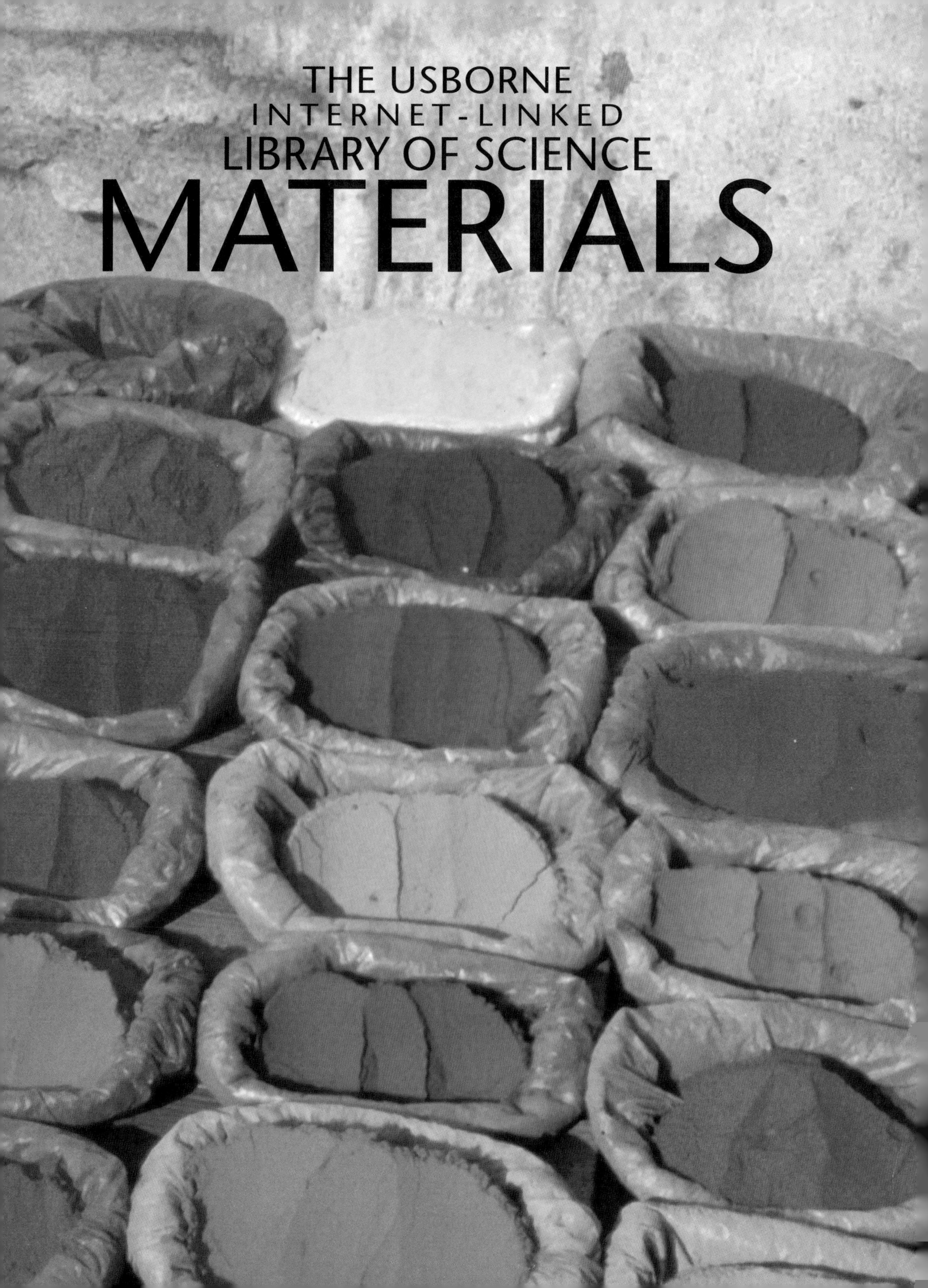

THE USBORNE
INTERNET-LINKED
LIBRARY OF SCIENCE

MATERIALS

Usborne Publishing has made every effort to ensure that material on the Web sites recommended in this book is suitable for its intended purpose. All the sites in this book have been selected by Usborne editors as suitable, in their opinion, for children, although no guarantees can be given. Usborne Publishing is not responsible for the accuracy or suitability of the information on any Web site other than its own. We recommend that young children are supervised while on the Internet, and that children do not use Internet chat rooms.

First published in 2001 by Usborne Publishing Ltd,
Usborne House, 83-85 Saffron Hill, London EC1N 8RT, England.

www.usborne.com

Printed in Spain

AE First published in America, 2002.

THE USBORNE
INTERNET-LINKED
LIBRARY OF SCIENCE

MATERIALS

Alastair Smith, Phillip Clarke and Corinne Henderson

Designed by Ruth Russell, Jane Rigby,
Chloë Rafferty and Adam Constantine

Digital illustrations by Verinder Bhachu
Digital imagery by Joanne Kirkby

Edited by Laura Howell

Cover design: Nicola Butler

Consultants: Elaine Wilson and Mark Beard

Web site adviser: Lisa Watts
Editorial assistant: Valerie Modd

Managing designer: Ruth Russell
Managing editor: Judy Tatchell

INTERNET LINKS

If you have access to the Internet, you can visit the Web sites we have recommended in this book. On every page, you will find descriptions of what is on each Web site, and why they are worth visiting. Here are some of the things you can do on the recommended sites in this book:

- watch animated movies about metals, atoms and other topics
- find out about the everyday uses of materials through videos, experiments and games
- browse galleries of amazing images from the natural world, including erupting geysers and volcanoes
- try simple experiments to create giant bubbles, build a carbon atom, and much more
- discover how to grow your own crystals

USBORNE QUICKLINKS

To visit the recommended sites in this book, go to the Usborne Quicklinks Web site, where you'll find links you can click on to take you straight to the sites. Just go to *www.usborne-quicklinks.com* and follow the simple instructions you find there.

Sometimes, Web addresses change or sites close down. We regularly review the sites listed in Quicklinks and update the links if necessary. We will provide suitable alternatives at *www.usborne-quicklinks.com* whenever possible. Occasionally, you may get a message saying that a site is unavailable. This may be a temporary problem, so try again later.

DOWNLOADABLE PICTURES

Pictures marked with the symbol ★ may be downloaded for your own personal use, for example, for homework or for a project, but may not be used for any commercial or profit-related purpose. To find these pictures, go to Usborne Quicklinks and follow the instructions there.

USING THE INTERNET

You can access most of the Web sites described in this book with a standard home computer and a Web browser (this is the software that enables you to access Web sites and view them on your computer).

Some Web sites need extra programs, called plug-ins, to play sounds or to show videos or animations. If you go to a site and you don't have the right plug-in, a message saying so will come up on the screen. There is usually a button you can click on to download the plug-in. Alternatively, go to Usborne Quicklinks and click on Net Help, where you will find links to plug-ins.

INTERNET SAFETY

Here are three important guidelines to follow to keep you safe while you are using the Internet:

- If a Web site asks you to register or log in, ask permission from your parent or guardian before typing in any information.
- Never give out personal information, such as your home address or phone number.
- Never arrange to meet someone that you communicated with on the Internet.

www.usborne-quicklinks.com

Go to Usborne Quicklinks for:

- direct links to all the Web sites described in this book
- free downloadable pictures, which appear throughout this book marked with a ★ symbol

SEE FOR YOURSELF

The *See for yourself* boxes in this book contain experiments, activities or observations which we have tested. Some recommended Web sites also contain experiments, but we have not tested all of these. This book will be used by readers of different ages and abilities, so it is important that you do not tackle an experiment on your own, either from the book or the Web, that involves equipment that you do not normally use, such as a kitchen knife or stove. Instead, ask an adult to help you.

CONTENTS

This metal was changed from a solid to a liquid by adding huge amounts of heat. In this form, it is described as molten.

MATERIALS

Everything in existence is made of tiny particles called atoms. Substances such as oxygen, which are made up of only one kind of atom, are called elements. Normally, atoms bond together to form groups called molecules, the smallest part of any material that can exist alone. In this book, you can learn about the chemistry of many different materials and the elements from which they are made.

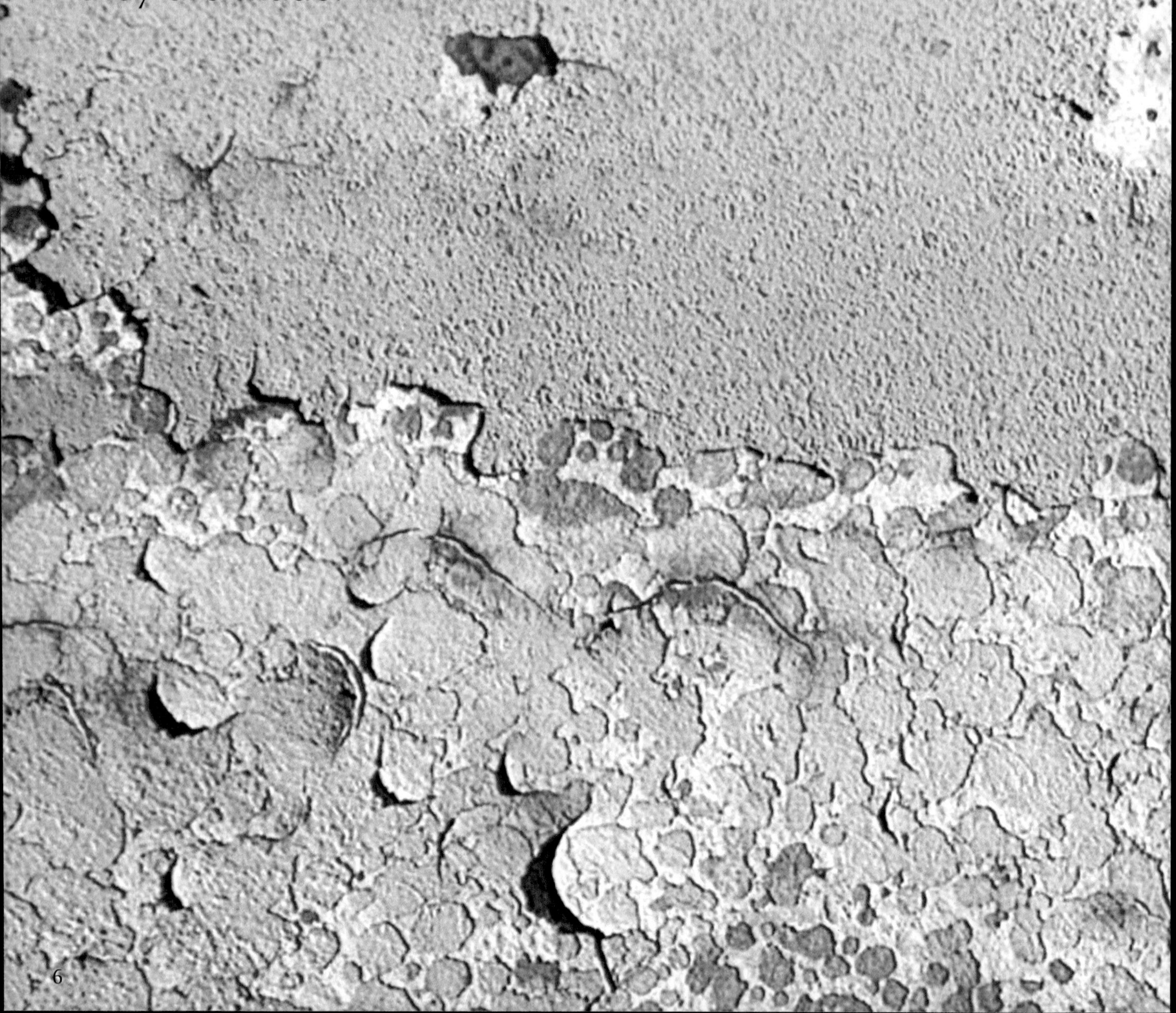

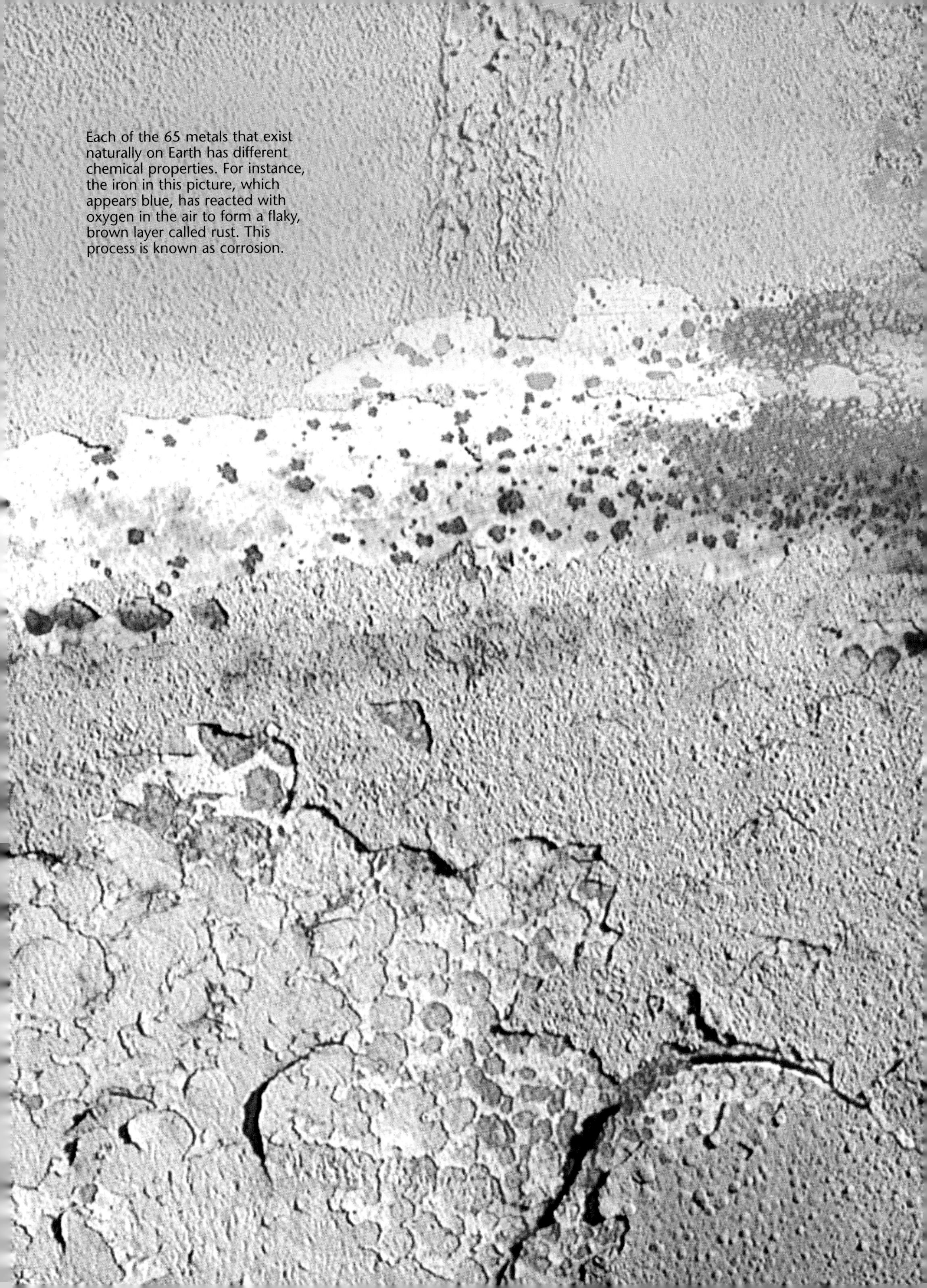

Each of the 65 metals that exist naturally on Earth has different chemical properties. For instance, the iron in this picture, which appears blue, has reacted with oxygen in the air to form a flaky, brown layer called rust. This process is known as corrosion.

ATOMIC STRUCTURE

Atoms are the tiny particles of which everything is made. It is impossible to imagine how small an atom is. A hundred million atoms side by side would measure only 1cm, and a sheet of paper, like the ones that make up this book, is probably a million atoms thick.

SUBATOMIC PARTICLES

Atoms are made of smaller particles called **subatomic particles**. In the middle of every atom is its **nucleus**. The nucleus contains two types of subatomic particles, called **protons** and **neutrons**.

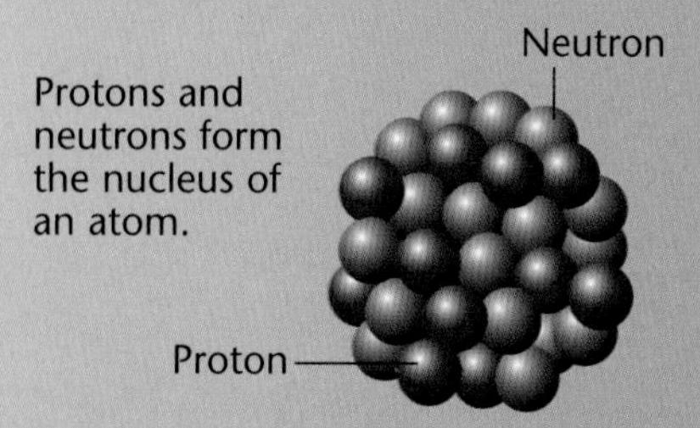

Protons and neutrons form the nucleus of an atom.

Subatomic particles of a third type, called **electrons**, move around the nucleus. The electrons exist at different energy levels, called **shells**, around the nucleus. Each shell can have up to a certain number of electrons. When it is full, a new shell is started.

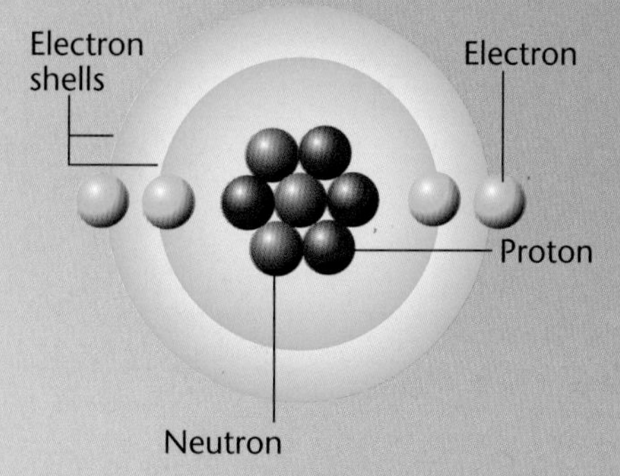

Scientists now think that protons and neutrons are made of even smaller subatomic particles, called **quarks**.

This diagram uses colored balls to represent the parts of an atom and illustrate the relationships between them.

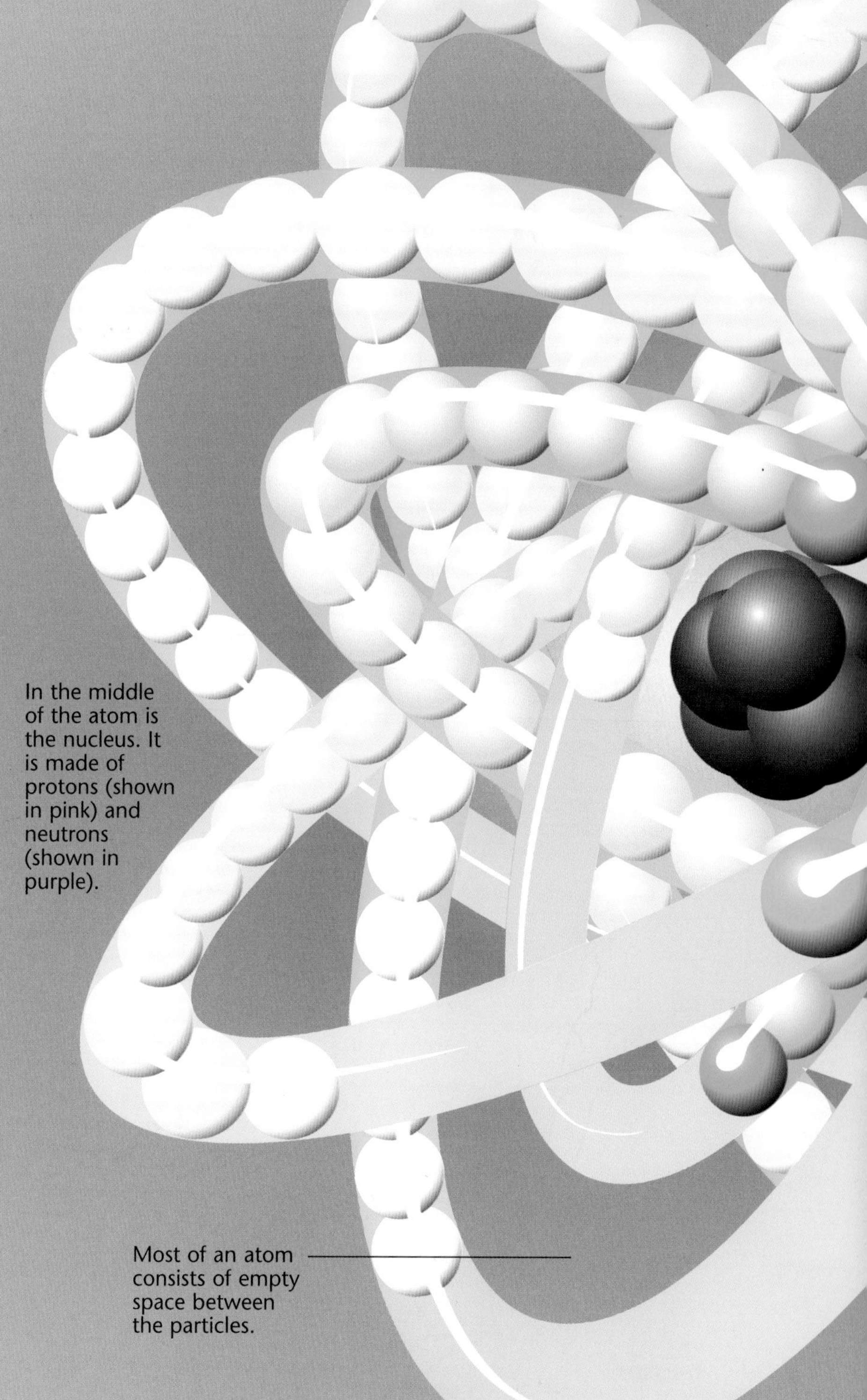

In the middle of the atom is the nucleus. It is made of protons (shown in pink) and neutrons (shown in purple).

Most of an atom consists of empty space between the particles.

Electrons are trapped by their attraction to protons, which are in the nucleus. They whizz around the nucleus at different levels, called shells.

The two electrons shown in green exist in the first shell of this atom. Those in blue are in the second shell.

Electron

ELECTRICAL CHARGES

The subatomic particles that make up an atom are held together by electrical charges. Particles with opposite electrical charges are attracted to one another.

The protons have a positive electrical charge and the electrons have a negative charge. Neutrons have no electrical charge, so they are neutral.

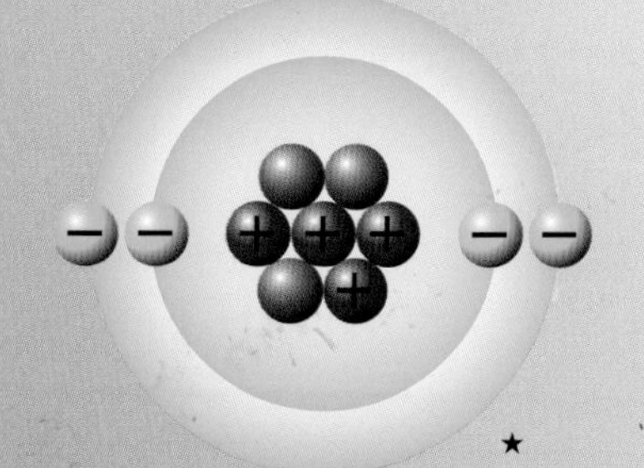

Proton: positive electrical charge.

Electron: negative electrical charge.

Neutron: no electrical charge.

An atom usually has an equal number of positively charged protons and negatively charged electrons. This makes the atom itself electrically neutral.

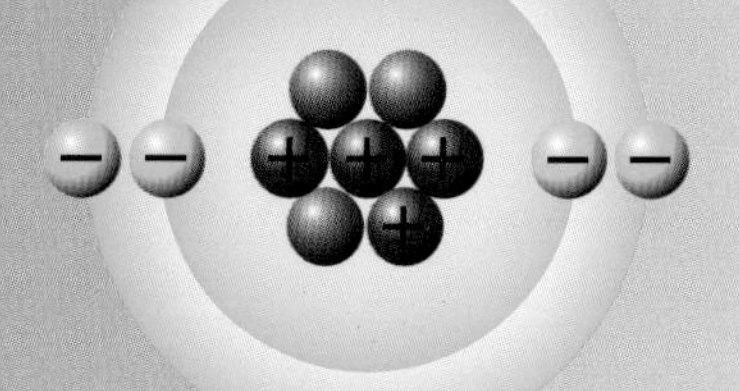

This atom is electrically neutral.

It has four protons.

It has four electrons.

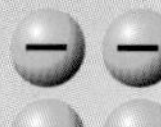

Its three neutrons have no effect on its electrical charge.

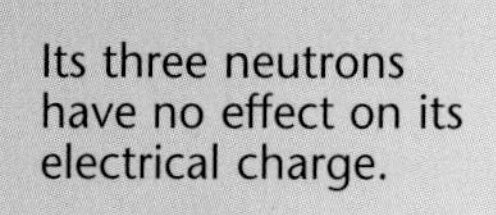

REPRESENTING ATOMS

Although atoms are often represented by diagrams like the main picture, scientists now believe that the electrons are held in cloud-like regions around the nucleus, as in the **electron cloud model** below.

Electron cloud model

Electrons can be anywhere within their cloud, at any time. Sometimes they even move outside it.

ELECTRON DENSITY

In the picture below, different colors show different levels of density of electrons in a group of atoms. The turquoise areas show where the electrons are most dense.

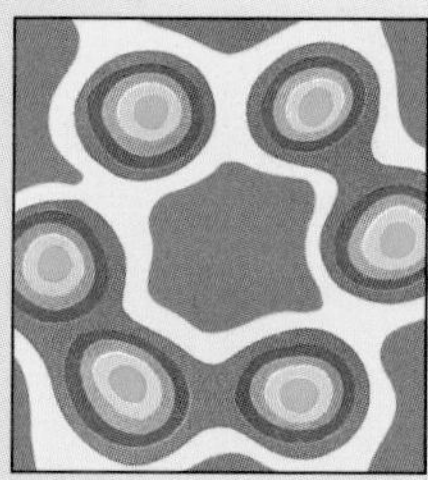

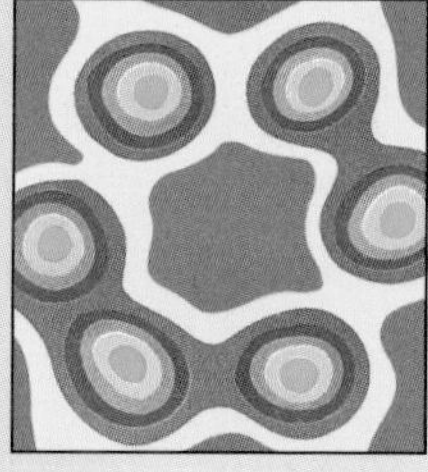

This is a picture of what you might see through an extremely powerful microscope.

Internet links

- Go to **www.usborne-quicklinks.com** for a link to the **BrainPop Web site** for a movie and quiz about atoms.
- Go to **www.usborne-quicklinks.com** for a link to the **Atoms Family Web site** to try the "Spectroscope of an Atom" and "Paper Cutting" activities in "The Phantom's Portrait Parlour".
- Go to **www.usborne-quicklinks.com** for a link to the **Particle Adventure Web site** for a guide to atoms and subatomic particles.
- Go to **www.usborne-quicklinks.com** for a link to the **Life, the Universe and the Electron Web site** for an online exhibition celebrating the centenary of the discovery of the electron.

ATOMIC NUMBER

Atoms of different substances have different numbers of protons in their nucleus. The number of protons in the nucleus is called the **atomic number**.

The atomic number of an atom indicates what substance it is.

An atom usually has an equal number of protons and electrons, so the atomic number also shows how many electrons it has.

Proton

Neutron

The nucleus of a carbon atom has six protons, so its atomic number is six.

The nucleus contains six protons and six neutrons, so its mass number is 12.

The nucleus of a phosphorus atom has 15 protons, so its atomic number is 15.

The nucleus contains 15 protons and 16 neutrons, so its mass number is 31.

MASS NUMBER

The more protons and neutrons an atom has, the greater its mass (the measurement of the amount of matter in the atom). The total number of protons and neutrons in an atom is called its **mass number**.

Electrons are left out of the mass calculation as they add so little to the mass of an atom.

A machine called a **mass spectrometer** can be used to help identify atoms by sorting them by mass.

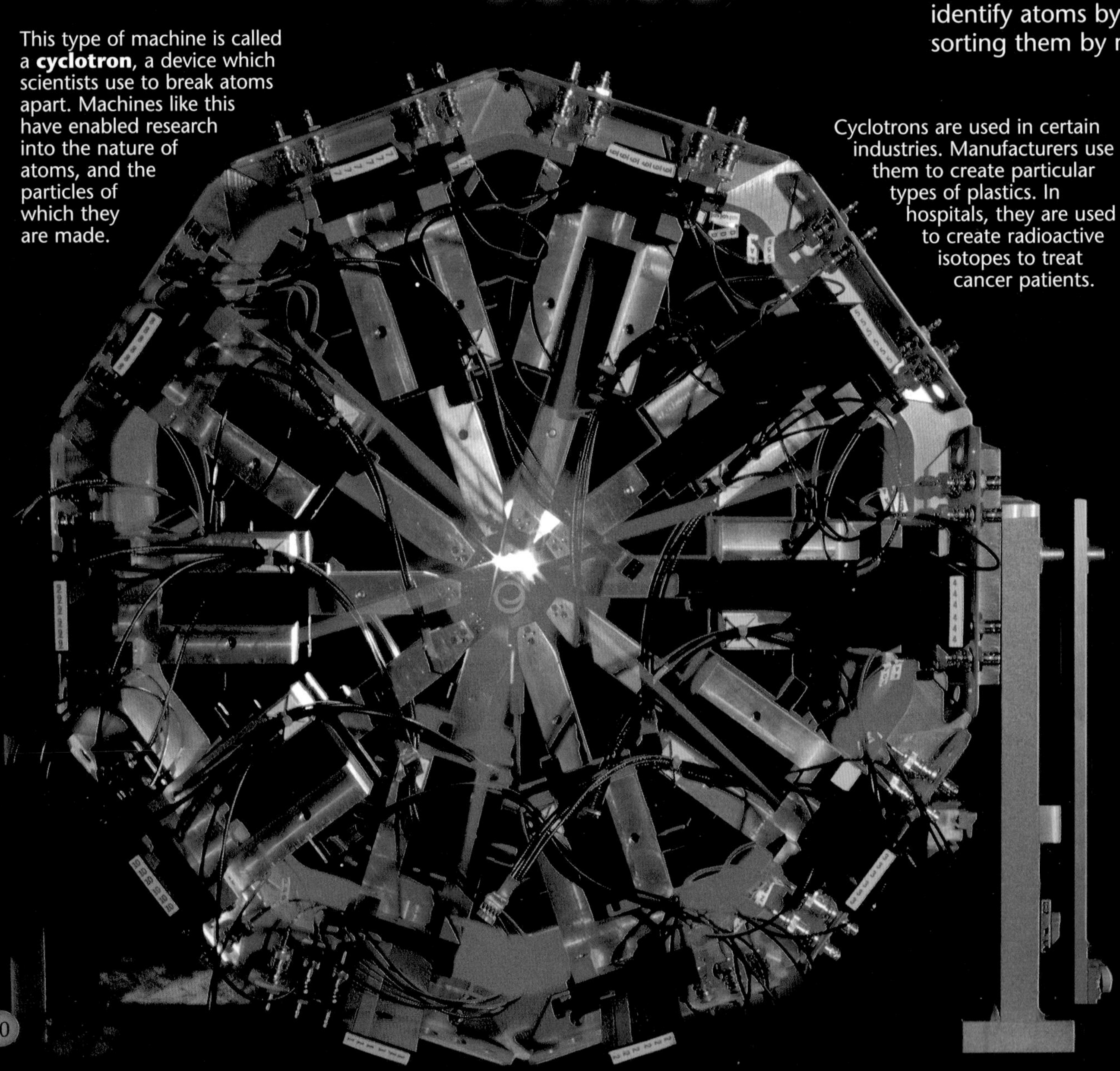

This type of machine is called a **cyclotron**, a device which scientists use to break atoms apart. Machines like this have enabled research into the nature of atoms, and the particles of which they are made.

Cyclotrons are used in certain industries. Manufacturers use them to create particular types of plastics. In hospitals, they are used to create radioactive isotopes to treat cancer patients.

ISOTOPES

Most atoms exist in a number of different forms, called **isotopes**. Each form has the same number of protons and electrons, but a different number of neutrons. So all the isotopes of an atom have the same atomic number, but they have different mass numbers.

The mass number of the isotope of an atom is written beside its name. For instance, carbon-12 has six protons and six neutrons.

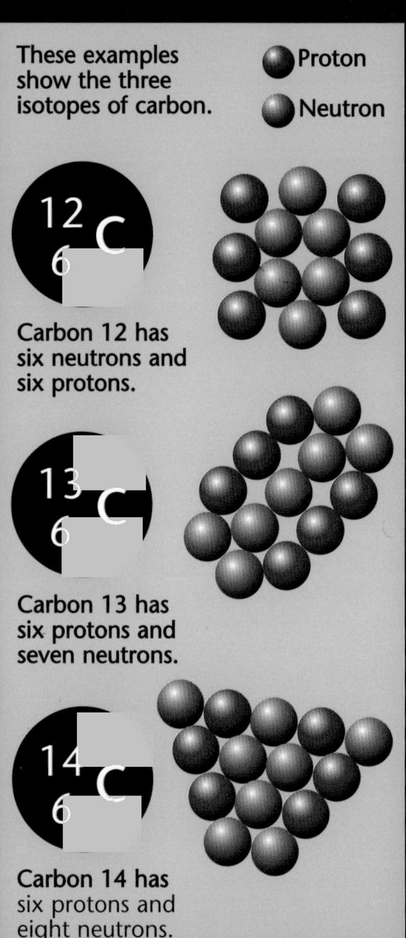

These examples show the three isotopes of carbon.
Proton
Neutron
12 C 6
Carbon 12 has six neutrons and six protons.
13 C 6
Carbon 13 has six protons and seven neutrons.
14 C 6
Carbon 14 has six protons and eight neutrons.

Isotopes have different physical properties but their chemical properties are the same. Most of the atoms in an **element** (a substance made up of only one type of atom) are a single isotope, with small amounts of other isotopes.

ANCIENT IDEAS

The idea that everything in the universe is made up of atoms is not a new one. Philosophers in Ancient Greece, 2,500 years ago, believed that matter was made up of particles that could not be cut any smaller. The word "atom" comes from the Greek word *atomos*, which means "uncuttable".

The theories of Aristotle, an Ancient Greek philosopher, influenced scientists and their studies on atoms for centuries.

Aristotle (384-322BC)

ATOMIC THEORY

The term "atom" was first used by the British chemist, John Dalton, when he put forward his **atomic theory** in 1807.

Dalton suggested that all chemical elements were made of very small particles, called atoms, that did not break up when chemicals reacted. He thought that every chemical reaction was the result of atoms joining or separating. Dalton's atomic theory provided the basis for modern science.

John Dalton (1766-1844)

Dalton used symbols to represent one atom of each element or substance.

Examples of Dalton's symbols

Zinc Mercury Sulfur

EARLY MODELS

Early in the twentieth century, scientists began to make models of atoms.

Ernest Rutherford (1871-1937) showed electrons with a negative electric charge circling a positively charged nucleus.

Rutherford's model

Niels Bohr (1885-1962) showed a model with electrons following specific orbits. In 1932, James Chadwick (1891-1974) showed the nucleus made of particles called neutrons and protons.

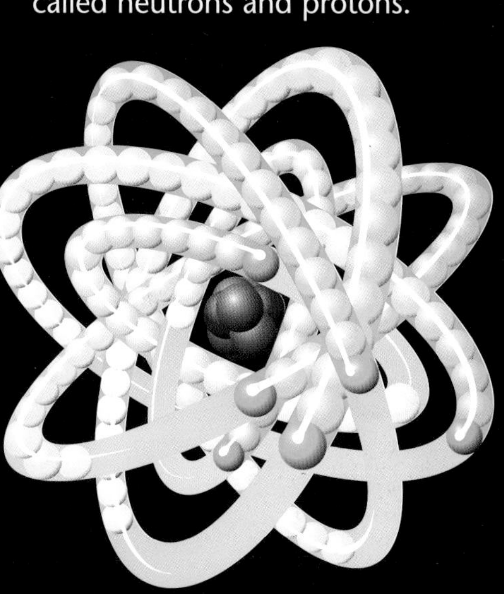

This model of an atom, which is also shown larger on pages 8-9, is based on models by Rutherford, Bohr and Chadwick.

Internet links

- Go to **www.usborne-quicklinks.com** for a link to the **Physics 2000 Web site** to find an introduction to elements by clicking on "Elements as Atoms".
- Go to **www.usborne-quicklinks.com** for a link to the **PBS Atom Builder Web site** to read the Atom Builder Guides, then build your own carbon atom.
- Go to **www.usborne-quicklinks.com** for a link to the **European Organization for Nuclear Research Web site** to explore a particle accelerator laboratory.

MOLECULES

Atoms are rarely found on their own. They usually cling, or bond, together to form molecules or large lattice structures. A **molecule** is a group of atoms that are bonded together to form the smallest piece of a substance that normally exists on its own. Molecules are much too small to be seen with the naked eye.

SHELLS AND BONDING

Most atoms have several shells of electrons*. The first shell of an atom can hold two electrons. The second and third shells can hold eight, although some atoms can hold up to 18 electrons in their third shells. When a shell is full, the electrons start a new shell. An atom is particularly stable when it has a full outer shell of electrons.

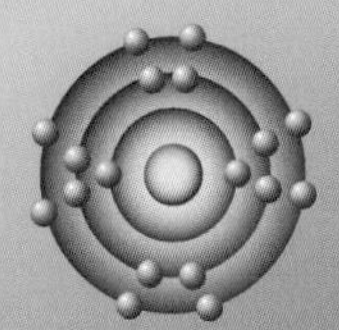
An argon atom has three full shells of electrons. It is a stable atom.

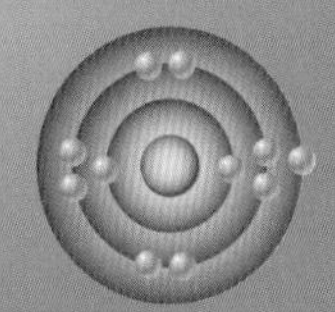
A sodium atom is unstable. It has only one electron in its third shell.

Atoms bond together in order to become stable. They do this by sharing electrons, or by giving up or taking electrons from another atom in order to achieve a full, or fuller, outer shell. Two hydrogen atoms, for instance, bond together to make a hydrogen molecule. They share their electrons, giving each atom a full outer shell.

Two hydrogen atoms

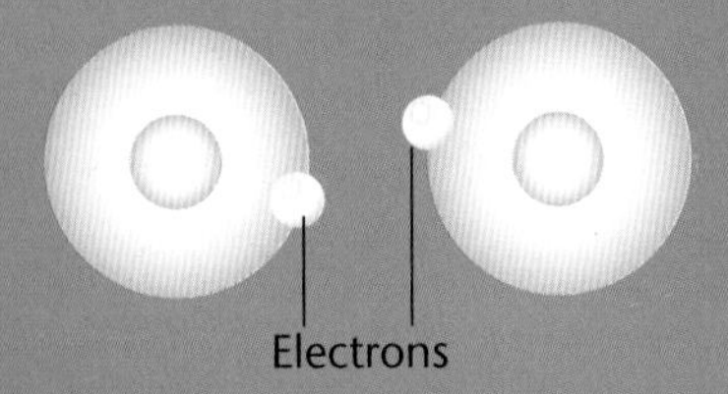

Hydrogen molecule

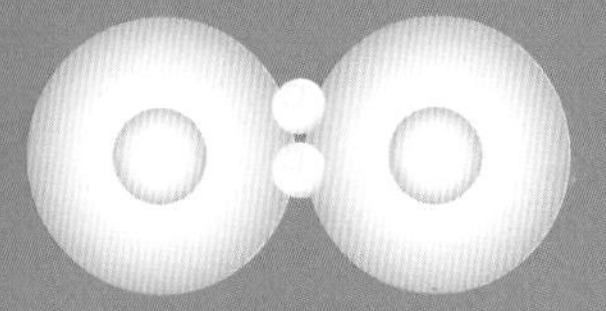
Each atom has a full shell of two electrons, so it is stable.

* Electrons, 8.

This is a model of a molecule of DNA, a complicated chemical compound found in the cells of all living things.

WATER MOLECULE

A water molecule consists of two different elements: hydrogen and oxygen. The two hydrogen atoms share electrons with the oxygen atom, so each has a complete shell. The oxygen atom uses two electrons (one from each hydrogen atom) to complete its own outer shell. All the atoms become stable.

The oxygen atom contains six electrons in its outer shell. It needs two more electrons to complete its outer shell and become stable.

Each hydrogen atom contains one electron in its shell. They each seek one electron to complete their shells and become stable.

CHEMICAL FORMULAE

An atom's name can be shown by a symbol (its **chemical symbol**). This is usually the first letter or two of its name in English, Latin or Arabic.

O	The symbol for oxygen
Au	The symbol for gold, from the Latin word *aurum*
Fe	The symbol for iron, from the Latin word *ferrum*
K	The symbol for potassium, from the Arabic word *kalium*

A **chemical formula** shows the atoms of which a substance is made and in what proportions. For example, each molecule of carbon dioxide is made up of one atom of carbon and two atoms of oxygen, so the formula for carbon dioxide is CO_2. The figure "2" shows the number of oxygen atoms in the molecule.

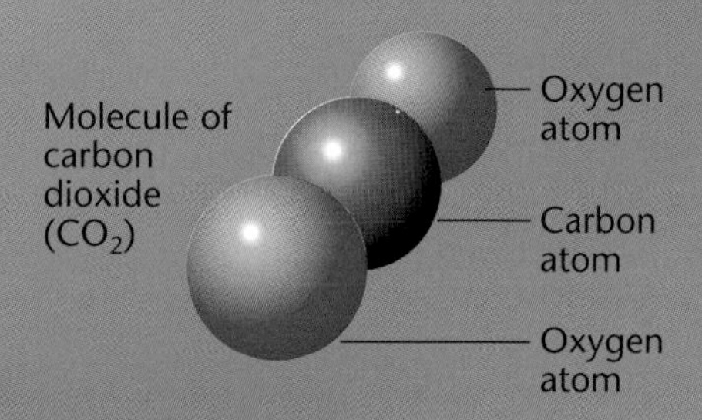

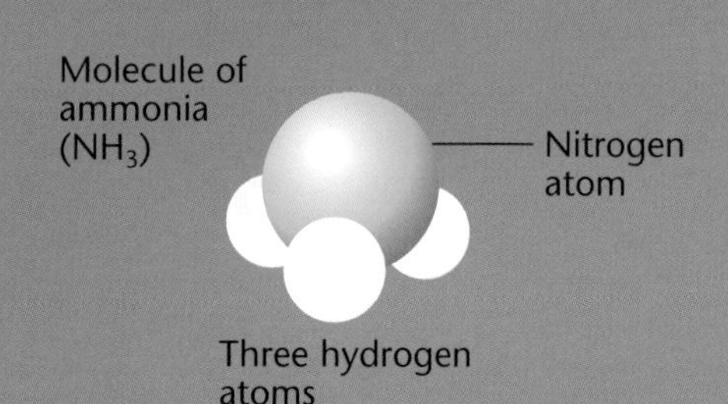

MODELS OF MOLECULES

When studying molecules, scientists often use models to represent them. There are two main types: ball-and-spoke models and space-filling models.

In **ball-and-spoke models**, the bonds that hold the atoms together are shown as sticks.

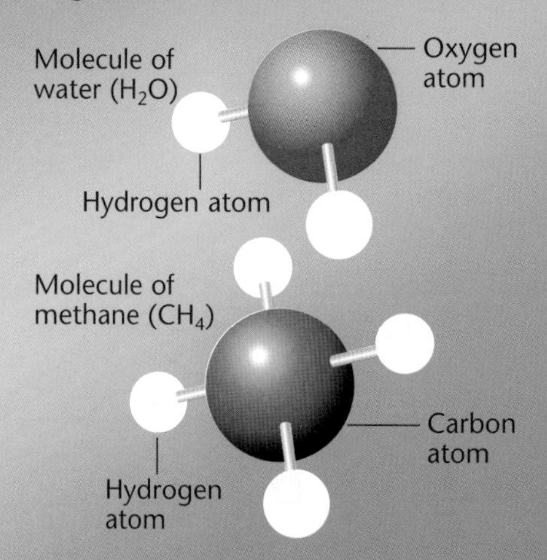

In **space-filling models**, atoms are shown clinging together.

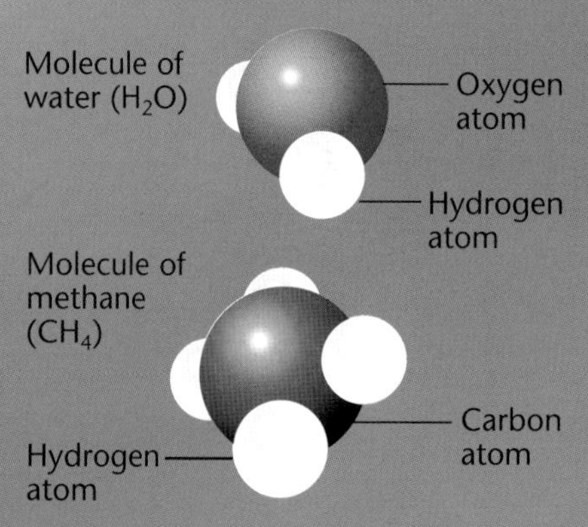

Neither model looks like an actual molecule, but they are simple ways of showing the atoms that form the molecule.

Internet links

- Go to **www.usborne-quicklinks.com** for a link to the **WebMolecules Web site** for a huge database of molecular models.
- Go to **www.usborne-quicklinks.com** for a link to the **Oxford University Web site** to browse through the "Molecules of the Month".
- Go to **www.usborne-quicklinks.com** for a link to the **IPPEX Web site** for an interactive lesson about molecules.

SOLIDS, LIQUIDS AND GASES

Most substances can exist in three different forms: as solids, liquids or gases. These are called the **states of matter**. A solid has a definite volume and shape. A liquid has a definite volume, but its shape changes according to the shape of its container. A gas has neither shape nor volume. It will move to fill the space available.

THE KINETIC THEORY

The theory that explains the properties of solids, liquids and gases is called the **kinetic theory**. It is based on the idea that all substances are made up of moving particles. It explains the properties of solids, liquids and gases in terms of the energy of these particles.

Heating a substance gives the particles more energy, enabling them to move around faster and change from one state to another. (See *Changes of State*, pages 16-17).

Like many scientific theories, the kinetic theory has never been proved. It provides an explanation, though, for how solids, liquids and gases are seen to behave, and why substances change from one state to another.

Movement of particles in solids, liquids and gases

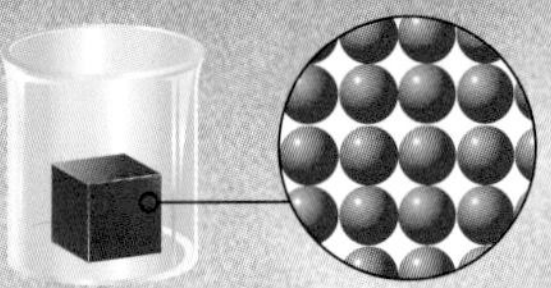

The particles in a solid have the least energy and cannot overcome the attraction between one another. They vibrate, but stay where they are.

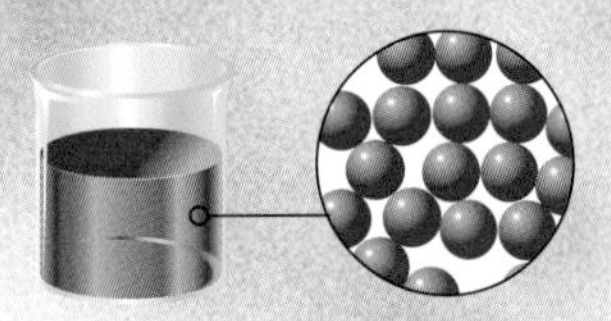

Heating a solid gives the particles more energy so they can escape from each other. The solid melts and becomes a liquid.

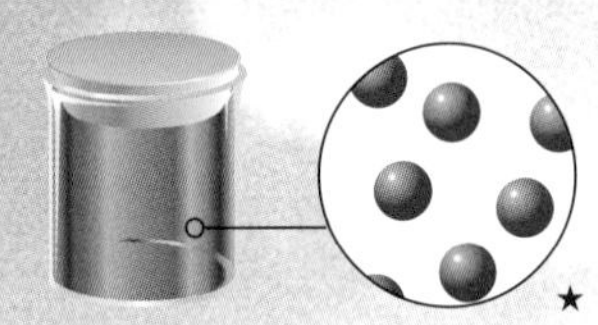

The particles in a gas have even more energy. They easily move far apart and spread out through the available space.

This is a geyser. Water heated to boiling point under the ground turns from a liquid to a gas (steam) and shoots out of a crack. You can find out more about what causes geysers over the next page.

BROWNIAN MOTION

Random movement of particles in a liquid

The movement of particles in liquids and gases is known as **Brownian motion**, named after a British biologist, Robert Brown (1773-1858). In 1827, Brown observed how tiny grains of pollen moved around randomly in a liquid, but he could not explain what caused this movement.

The German-born scientist Albert Einstein (1879-1955) later explained that the movement of particles in a liquid or gas is caused by the particles being hit by the invisible molecules of the fluid in which the particles are floating.

MEASURING SUBSTANCES

Volume is the amount of space occupied by a solid or liquid. It is measured in cubic meters (m^3).

You can calculate the volume of a rectangular solid using this formula:

Volume = Length x Breadth x Height

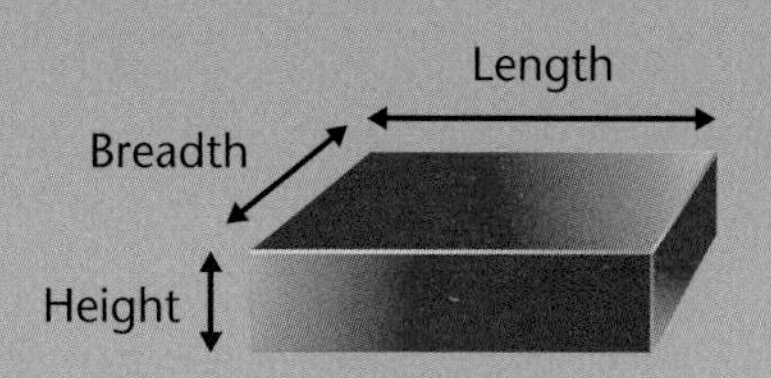

The volume of a liquid can be found by pouring the liquid into a measuring cylinder marked with a scale.

Measuring cylinder

The volume of an irregularly shaped solid is measured by finding how much liquid it displaces, using a Eureka can.

Eureka can

1. Eureka can is filled with water to base of spout.

2. Object is put into Eureka can.

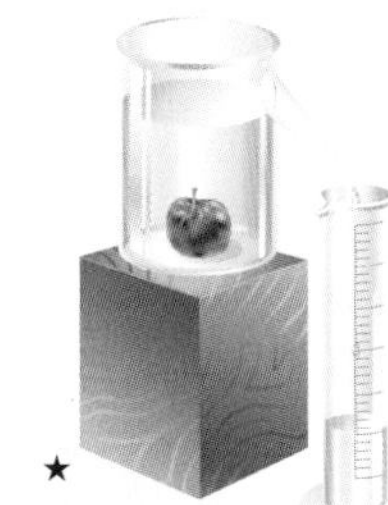

3. Volume of displaced water is measured.

The **mass** of a solid, liquid or gas is the amount of matter it contains. This is measured in kilograms. Mass is different from **weight**, which is a measure of the strength of the pull of gravity on an object. Mass is measured by weighing a substance and comparing its mass with a known mass.

Density is the mass of a substance compared with its volume. For example, the same volumes of cork and metal have different densities because the mass of the metal is much greater than that of the cork. Density is found by dividing the mass of an object by its volume, and it is measured in kilograms per cubic meter (kg/m^3).

$$\textbf{Density} = \frac{\textbf{Mass}}{\textbf{Volume}}$$

The density of a liquid is measured using a **hydrometer**. The hydrometer floats near the surface in a dense liquid, as only a small volume of liquid needs to be displaced to equal the weight of the hydrometer.

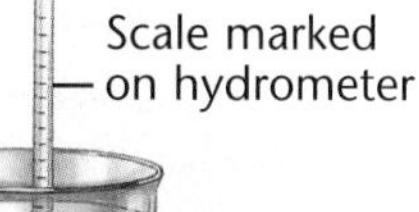

See for yourself

You can do an experiment to find the volumes of irregularly shaped solids without using a Eureka can.

You will need a measuring cup, a cake mixing bowl and a dishpan.

First, put your mixing bowl inside the dishpan. Then carefully fill the mixing bowl with water up to the brim.

Now take the object that you want to measure and hold it just on top of the water's surface. Let the object sink into the water. Water will slop over the side of the mixing bowl and be caught in the dishpan.

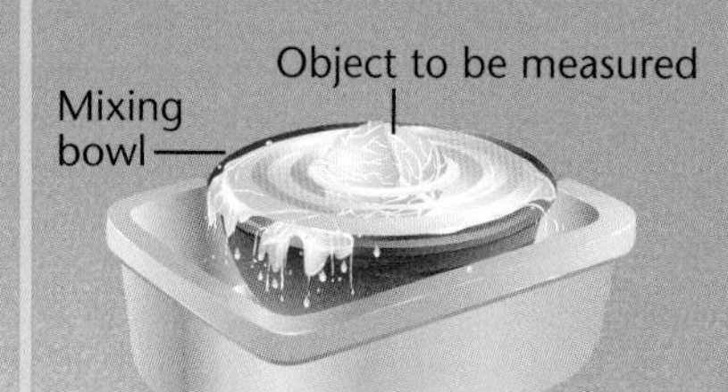

Take the mixing bowl out of the dishpan. Now pour the water from the dishpan into the measuring cup. The volume of water is equal to the volume of the object.

Internet links

- Go to **www.usborne-quicklinks.com** for a link to the **Chem4Kids Web site** for an introduction to the states of matter.
- Go to **www.usborne-quicklinks.com** for a link to the **ChangChem Web site** to see animations and explanations by clicking on "The Kinetic Theory of Matter".
- Go to **www.usborne-quicklinks.com** for a link to the **Harcourt School Publishers Web site** for an amazing animation of how particles behave in a solid, a liquid and a gas.
- Go to **www.usborne-quicklinks.com** for a link to the **Exploratorium Web site** to discover how to make an oozy substance that can't decide if it's a solid or a liquid.
- Go to **www.usborne-quicklinks.com** for a link to the **IPPEX Web site** for an online lesson about the states of matter.

CHANGES OF STATE

A substance changes from one **state of matter**, that is solid, liquid or gas, to another, depending on its temperature and pressure. When something changes state, heat is produced or lost as the energy of its particles is increased or decreased. Different substances change state at different temperatures.

Ice cream melts and becomes a liquid in the heat of the Sun.

The heat from a flame melts candle wax, but the wax sets as it drips away from the flame and cools.

MELTING AND BOILING

When a solid is heated, its temperature rises and its particles gain energy until it reaches its **melting point**. The particles now have enough energy to break away from their neighbors so the solid melts.

Further heat causes the temperature of the liquid to rise until it reaches its **boiling point** and the particles break free of each other completely. The liquid becomes a gas.

Some substances, for example carbon dioxide, change from gas to solid, or solid to gas, without passing through a liquid form. This is called **sublimation**.

The temperature at which a substance melts or boils changes if it contains traces of any other substances. For instance, ice (the solid form of water) melts at 0°C. Adding salt to the ice lowers its melting point.

This popsicle melts at a lower temperature than pure water ice because orange juice has been added to it.

When steam cools down, it turns back into water.

GEYSERS

Geysers are jets of boiling hot water and steam that shoot out from the Earth's crust.

They occur when water under the ground is heated by hot rocks and begins to boil.

As the water turns to steam, the pressure builds up in the channels between the rocks. The geyser then erupts, shooting a jet of steam and water high up into the air.

How geysers occur

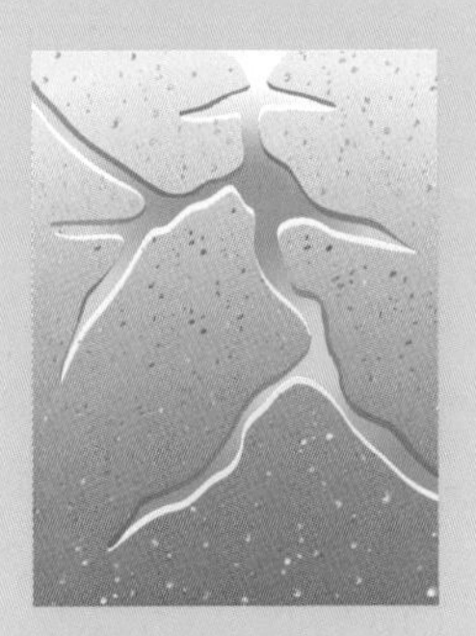

Water flows into cavities between the rocks under the ground.

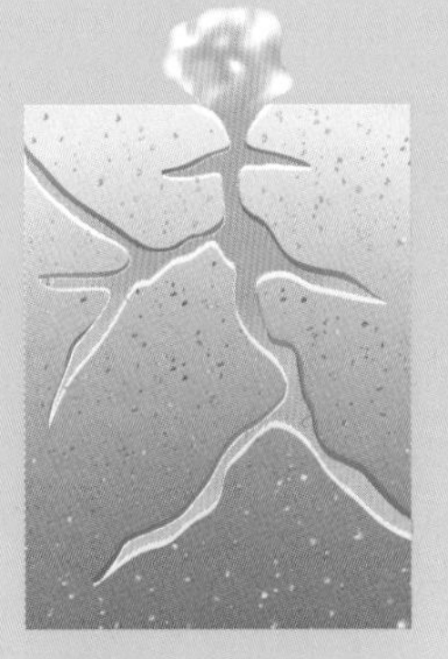

Pressure builds as the water heats and expands. Eventually, it turns to steam.

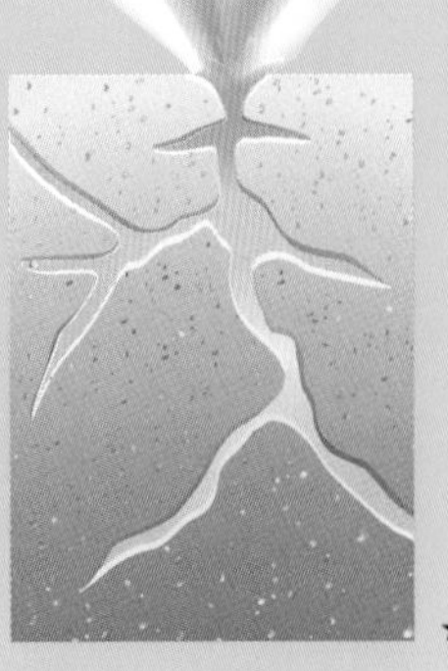

The pressure builds until boiling water and steam shoot out of a crack in the ground.

CONDENSATION

When a gas cools down enough, it **condenses**, becoming a liquid. This is because as it cools down, its particles lose energy and are unable to stay as far away from each other.

Condensation

Water vapor in the air in a room condenses on a cold window. Droplets of water are formed on the inside of the window.

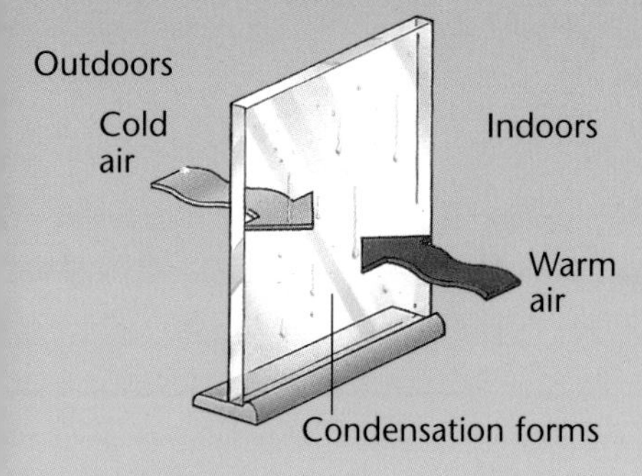

FREEZING

When a liquid cools enough, it sets or **freezes**, becoming a solid. Its particles lose further energy and are unable to overcome the attraction between each other.

When tiny droplets of water in the atmosphere freeze, they sometimes join together in beautiful patterns of crystals and form snowflakes like these.

PRESSURE

Air pressure has an effect on the melting or boiling point of a substance. The air naturally presses down on the Earth with a force called **atmospheric pressure**. At sea level, this is described as **one atmosphere**, or **standard pressure**.

At sea level, pure water boils at 100°C.

Higher up, the atmospheric pressure is less. It is easier for the particles in liquids to escape into the air, so their boiling points are lower.

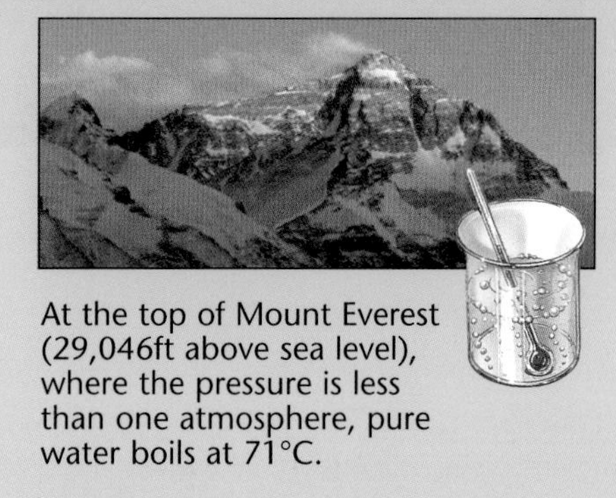

At the top of Mount Everest (29,046ft above sea level), where the pressure is less than one atmosphere, pure water boils at 71°C.

WATERLESS PLANET

The surface of Mars is dry. Scientists think that this is because the atmospheric pressure is very low, so any water immediately boils away.

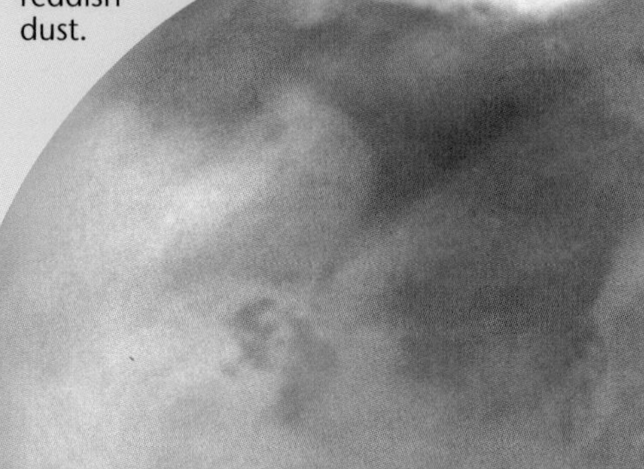

Mars is covered by a dry, reddish dust.

SOLID LIQUID OR GAS?

Whether something is classified as a solid, liquid or gas depends on its state at room temperature (20°C).

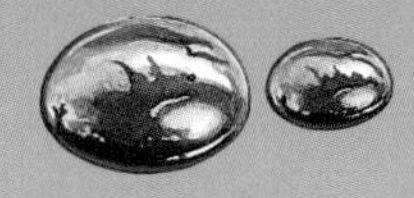

Mercury melts at -40°C. It is a liquid at room temperature.

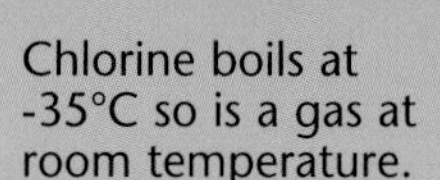

Chlorine boils at -35°C so is a gas at room temperature.

See for yourself

Fill a metal container with ice cubes. Stand it in a warm place and leave it for a few minutes. Then look at the container. You will see drops of water on the outside of it.

Water molecules in the warm air lose energy and slow down when they are cooled by the ice. They stick to each other, forming water droplets.

Droplets of water on the side of the can

Internet links

- Go to **www.usborne-quicklinks.com** for a link to the **How Stuff Works Web site** to learn how ice-rinks are kept frozen.
- Go to **www.usborne-quicklinks.com** for a link to the **Snow Crystals Web site** to find information on snowflakes.
- Go to **www.usborne-quicklinks.com** for a link to the **Yellowstone National Park Web site** for pictures of geysers.
- Go to **www.usborne-quicklinks.com** for a link to the **General Chemistry Online Web site** to discover why salt melts ice.
- Go to **www.usborne-quicklinks.com** for a link to the **Atoms Family Web site** to visit the "Molecule Chamber".

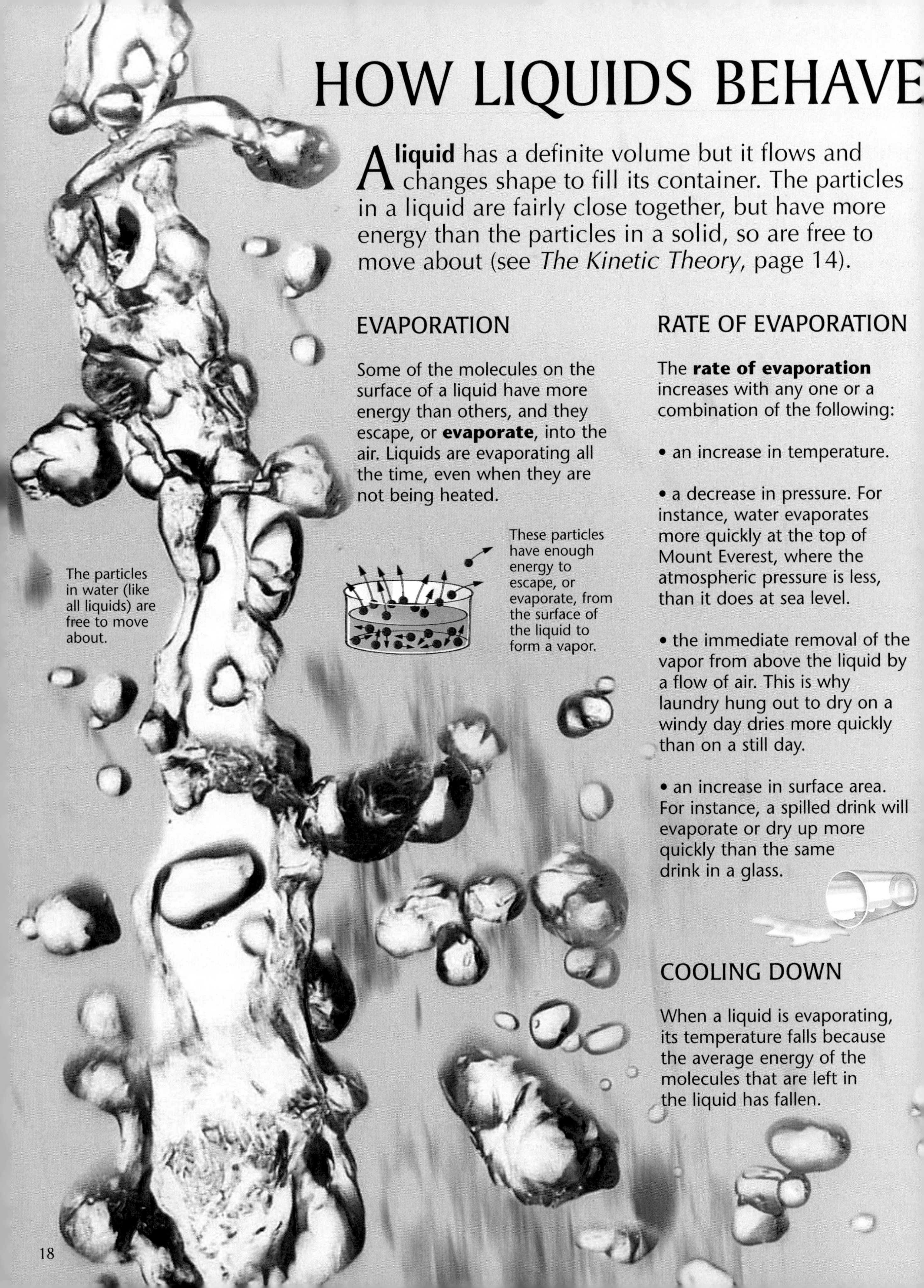

HOW LIQUIDS BEHAVE

A **liquid** has a definite volume but it flows and changes shape to fill its container. The particles in a liquid are fairly close together, but have more energy than the particles in a solid, so are free to move about (see *The Kinetic Theory*, page 14).

EVAPORATION

Some of the molecules on the surface of a liquid have more energy than others, and they escape, or **evaporate**, into the air. Liquids are evaporating all the time, even when they are not being heated.

The particles in water (like all liquids) are free to move about.

These particles have enough energy to escape, or evaporate, from the surface of the liquid to form a vapor.

RATE OF EVAPORATION

The **rate of evaporation** increases with any one or a combination of the following:

- an increase in temperature.

- a decrease in pressure. For instance, water evaporates more quickly at the top of Mount Everest, where the atmospheric pressure is less, than it does at sea level.

- the immediate removal of the vapor from above the liquid by a flow of air. This is why laundry hung out to dry on a windy day dries more quickly than on a still day.

- an increase in surface area. For instance, a spilled drink will evaporate or dry up more quickly than the same drink in a glass.

COOLING DOWN

When a liquid is evaporating, its temperature falls because the average energy of the molecules that are left in the liquid has fallen.

SURFACE TENSION

The molecules in a liquid are attracted by all the other liquid molecules around them. The ones on the surface, though, are not pulled upward because there are no liquid molecules above them. They are more attracted to the other liquid molecules than to the air.

This sideways and downward attraction at the surface creates a force called **surface tension**. It makes a liquid seem to have a "skin".

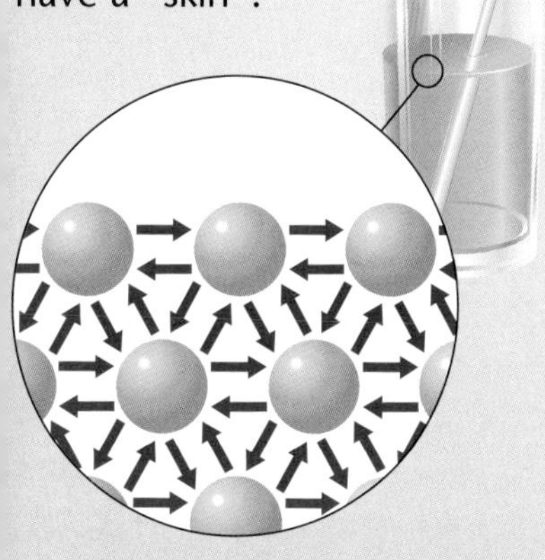

Molecules at the surface are attracted to each other, and to those below them. This creates surface tension.

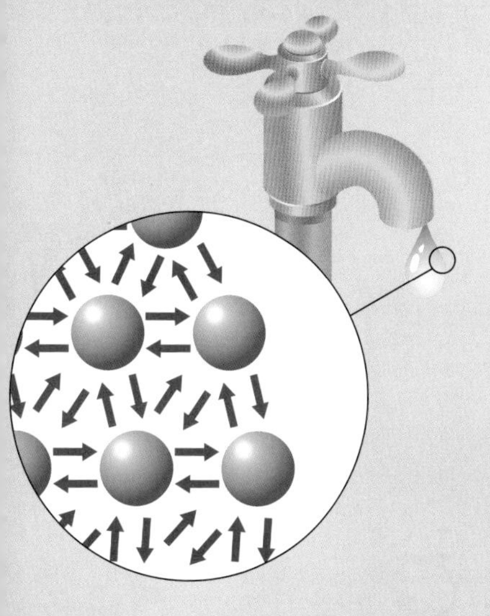

Water forms into droplets because surface tension pulls inward from all sides, keeping the molecules together.

The drips on these leaves form because surface tension pulls molecules of rainwater together.

STRETCHY SKIN

As a result of surface tension, a liquid's surface is like a stretchy skin, strong enough to support very light objects, such as dust or even insects.

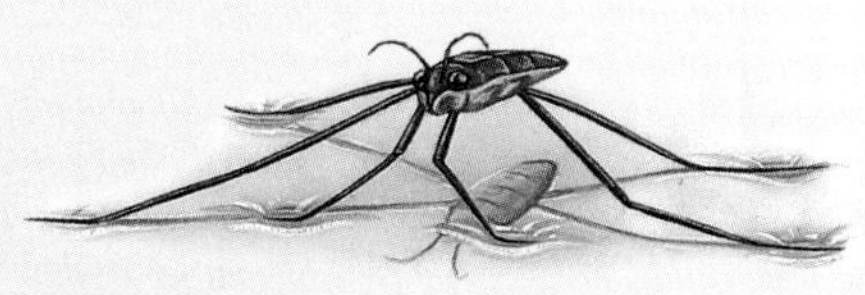

Pond skaters can walk on the surface of water as they are not heavy enough to break the skin-like surface tension.

COHESION

Cohesion happens when molecules of one substance are more attracted to each other than to a substance they are touching. Surface tension is an example of this. Molecules at the water's surface try to stay together rather than move toward the air above.

ADHESION

When molecules of a liquid are more attracted to a substance they are touching than to each other, **adhesion** occurs. The liquid adheres (sticks) to the other substance. Water does this when it touches the sides of a glass.

See for yourself

To see how surface tension can support certain objects, try this quick activity.

Fill a container with water. Put a needle on a small piece of tissue paper and lay it gently on the water.

The tissue soon becomes waterlogged and sinks, but the needle stays afloat, supported by surface tension.

Look closely and you will see that the needle actually dents the water's surface.

Internet links

- Go to **www.usborne-quicklinks.com** for a link to the **Exploratorium Web site** to find out how to create giant bubbles.
- Go to **www.usborne-quicklinks.com** for a link to the **Franklin Institute Online Web site** to investigate the power of soap and do some experiments.
- Go to **www.usborne-quicklinks.com** for a link to the **Exploratorium Web site** to discover if hot and cold liquids mix.
- Go to **www.usborne-quicklinks.com** for a link to the **Chem4Kids Web site** for easy-to-read information about liquids.

HOW GASES BEHAVE

A **gas** is a substance that has no definite volume or shape. Its particles have enough energy to spread far apart from each other and fill the space available.

Smells, such as the scent of flowers, are gases that travel through the air by diffusion.

DIFFUSION

The molecules in a gas have enough energy to break free of the forces between them (see *The Kinetic Theory*, page 14). They spread out to fill the available space. This is called **diffusion**.

During diffusion, molecules move from an area where they are in higher concentration to one where their concentration is lower. Diffusion stops when the molecules are evenly distributed.

Molecules of a light gas

Molecules of a heavy gas

Molecules of these two gases diffuse together over time. Light gases diffuse faster than heavy ones.

This scientist is taking samples of gases emerging from holes in the side of a volcano. Some of the gases are harmful, so the scientist has to wear a breathing mask.

The set-up below shows how two gases mix by diffusion. A jar of air is turned upside down on top of a jar of bromine, which is heavier than air.

After fifteen minutes, the air and bromine in the jars become mixed by diffusion as their molecules spread through the two jars.

PRESSURE, TEMPERATURE AND VOLUME

Gases exert a push on things that they are contained in. This push, called **pressure**, is felt in all directions. It is the rate at which molecules in a gas hit the sides of its container.

Any change in pressure, temperature, or the container's volume will cause a change in the molecules' behavior.

If the volume of a gas at a constant temperature is decreased, for example by reducing the size of its container, the pressure of the gas increases. This is because the gas molecules hit the walls of the container more frequently.

When heated, the molecules in a gas gain energy, move around faster and become even further apart – the gas expands and becomes less dense. This is why hot-air balloons float – the air inside them is less dense than the air around them.

If a gas is heated but is not allowed to expand, then its pressure increases. This is because the molecules in the gas gain energy, move around more quickly and hit the walls of the container more frequently.

Thermometer, to measure temperature.

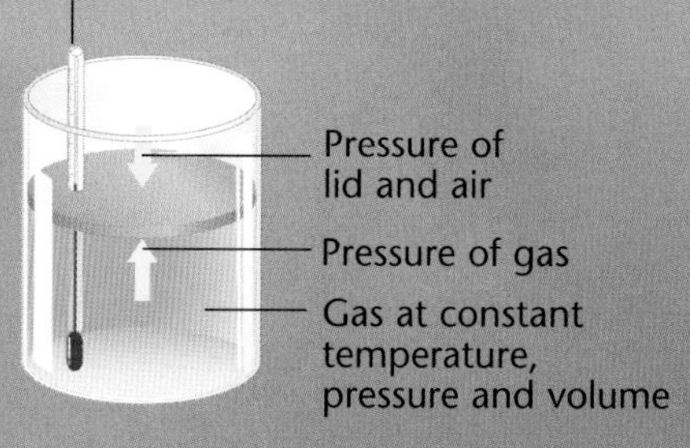
Pressure of lid and air
Pressure of gas
Gas at constant temperature, pressure and volume

Temperature is the same as before.

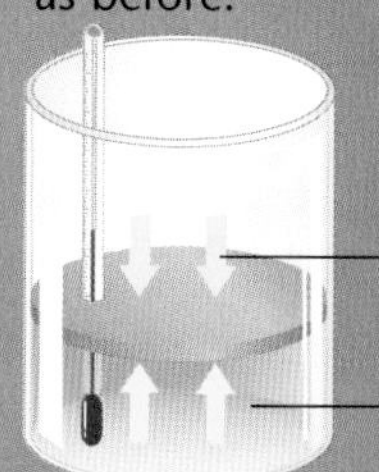
Pressure increased to reduce size of container.
Volume decreases, pressure increases.

Temperature increased.

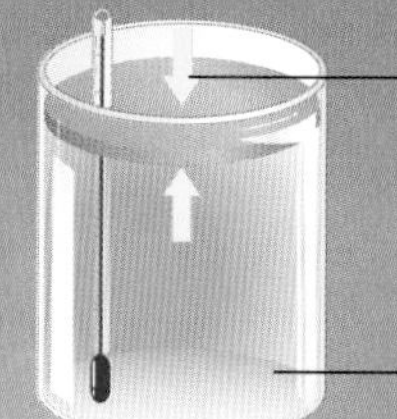
Pressure kept the same as before.
Volume increases until pressures are equal.

Temperature increased.

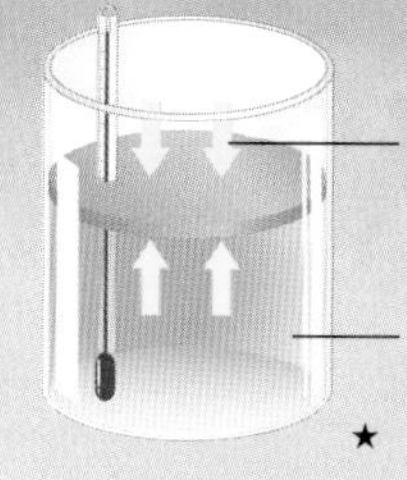
Pressure increased, to keep volume same as before.
Pressure increases.

Under the ground, volcanic gases become extremely hot. The pressure builds and builds until they shoot out of cracks and holes in the ground.

Balloons stretch as air spreads out to fill them.

See for yourself

Next time you use a balloon pump, see how it uses pressure to fill the balloon with air.

1. When you pump the handle, the volume of the pump's chamber is decreased so the air pressure inside it is increased.

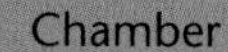
Chamber

2. Air shoots out of the nozzle, into the balloon.

3. A valve in the pump prevents air from being sucked back out of the balloon.

4. Pressure inside the balloon increases so it stretches and expands. Its volume increases.

Internet links

- Go to **www.usborne-quicklinks.com** for a link to the **Bubblesphere Web site** to investigate the properties of bubbles.
- Go to **www.usborne-quicklinks.com** for a link to the **BBC Education Web site** to find out how air pressure forces liquid up a straw.
- Go to **www.usborne-quicklinks.com** for a link to the **How Stuff Works Web site** to learn about smells.
- Go to **www.usborne-quicklinks.com** for a link to the **Hyper Chemistry Web site** for an experiment with gas.
- Go to **www.usborne-quicklinks.com** for a link to the **How Stuff Works Web site** to learn how gas is involved in making a refrigerator work.

THE ELEMENTS

An **element** is a substance that contains only one kind of atom – the tiny particles of which all substances are made. For example, sulfur, helium and iron are elements: they contain only sulfur, helium or iron atoms and they cannot be broken down into simpler substances.

GROUPING ELEMENTS

So far, 115 elements have been discovered, but only 90 occur naturally on Earth. Elements can be sorted into metals, non-metals and semi-metals and arranged in a table, called the periodic table, which is shown on pages 26-27.

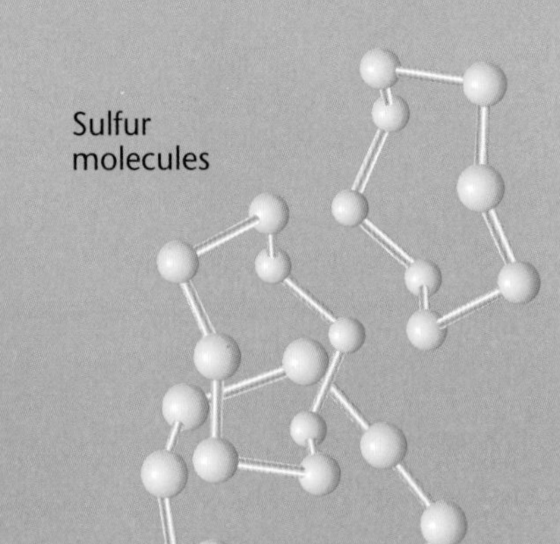

Sulfur is one of the 90 elements that occur naturally on Earth. It is a non-metal. Its molecules, as shown in this diagram, form irregular ring shapes consisting of eight sulfur atoms.

METAL ELEMENTS

Over three-quarters of all the elements are **metals**. Most of the metal elements are dense and shiny. They have many uses as they are strong, but can be easily shaped. They are also good conductors of heat and electricity. Metals are usually found combined with other elements in the Earth's crust (see pages 24-25).

These chocolate eggs are wrapped in thin aluminum foil to keep them fresh. Aluminum is the most common metal on Earth.

Here, aluminum is rolled into a long, thin sheet. It can be re-shaped easily without breaking because its atoms, which are closely packed, slide over each other.

The Space Shuttle relies on burning elements to blast it into space. It burns the non-metal hydrogen (stored in the red-brown colored external fuel tank) and powdered aluminum metal (stored in the two white rockets).

NON-METALS

There are 16 naturally occurring **non-metal** elements. All (apart from graphite, a form of carbon) are insulators – poor conductors of heat and electricity.

At room temperature, four non-metals (phosphorus, carbon, sulfur and iodine) are solids, and bromine is a liquid. The other 11 non-metals are gases.

Non-metals

Hydrogen	Sulfur
Helium	Chlorine
Carbon	Argon
Nitrogen	Bromine
Oxygen	Krypton
Fluorine	Iodine
Neon	Xenon
Phosphorus	Radon

SEMI-METALS

Semi-metals, also called **metalloids**, can act as poor conductors, just like non-metals. They can also be made to conduct well, like metals. Because of this, semi-metal elements are called **semiconductors**. There are nine semi-metals (see list, right). They are all solids at room temperature.

Semi-metals

Boron	Antimony
Silicon	Tellurium
Germanium	Polonium
Arsenic	Astatine
Selenium	

The semi-metal germanium is used to make transistors* like this one. They are used in radios.

Silicon is used to make integrated circuits* such as this one. Microscopic pathways in the circuit conduct and block electrical pulses.

See for yourself

Finding out how well a substance conducts heat can help to identify whether it is a metal or a non-metal. Try the experiment below.

You will need several long objects such as a metal spoon, a wooden spoon and a plastic ruler. Put a smear of cold butter near the end of each object.

Place the objects in a mug filled with warm water.

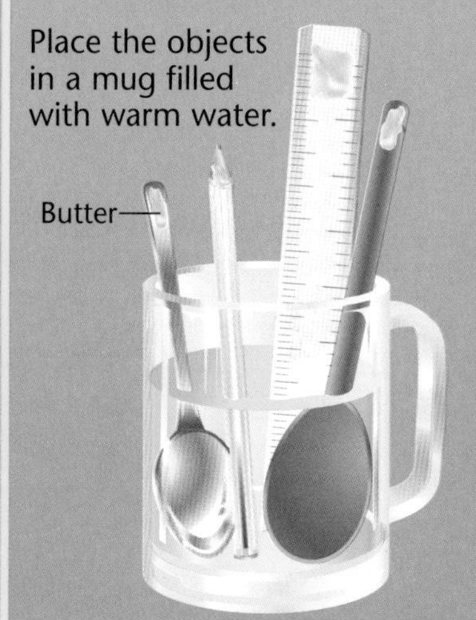

As the heat travels up the object, it melts the butter. You should find that the butter melts on the metal things first, because metals are better conductors of heat than non-metals. Eventually, the warmth of the rising air melts the butter on all the objects.

Internet links

- Go to **www.usborne-quicklinks.com** for a link to the **Periodic Chart Web pages** where you can read in-depth information on the elements.
- Go to **www.usborne-quicklinks.com** for a link to the **QUIA Web site** where you can play interactive games to learn all about the elements and their chemical symbols.
- Go to **www.usborne-quicklinks.com** for a link to the **Amethyst Galleries Web site** to discover illustrated facts about many different elements.
- Go to **www.usborne-quicklinks.com** for a link to the **Chemical Elements Web site** where you can find detailed information about metalloids and non-metals, then select other element groups from a list on the left of the screen.

* *Integrated circuit, 59; Transistor, 60.*

ELEMENTS IN THE EARTH

The outermost layer of the Earth is called the **crust**. Most of it is made of only five elements. It is rare for these elements to occur alone, though some, like gold, do. More often they are found together as combined substances called **compounds**. The pure and combined elements found in the crust are called **minerals**. Minerals that contain metals are called **ores**.

Some minerals, such as this chalcedony, can be polished to make beautiful decorative objects.

COMMON ELEMENTS

Oxygen is the most common element in the Earth's crust. It often occurs combined with silicon, the second most common element, and with aluminum and iron, the most common metals.

This pie chart shows the proportions, by mass, of the five main elements in the Earth's crust.

- Oxygen 46.6%
- Silicon 27.7%
- Aluminum 8.1%
- Iron 5%
- Calcium 3.6%
- Others 9%

MINERAL FORMATION

Most minerals are formed when hot **magma** (molten rock that contains dissolved gases) pushes up from deep below the Earth's crust, cools and solidifies.

The conditions in the place where magma cools determine which type of mineral forms. Geometric shapes called **crystals** form when minerals cool slowly. The cooling process can be so quick, though, that the mineral has no time to crystallize. A kind of shiny black glass, called obsidian, forms in these conditions.

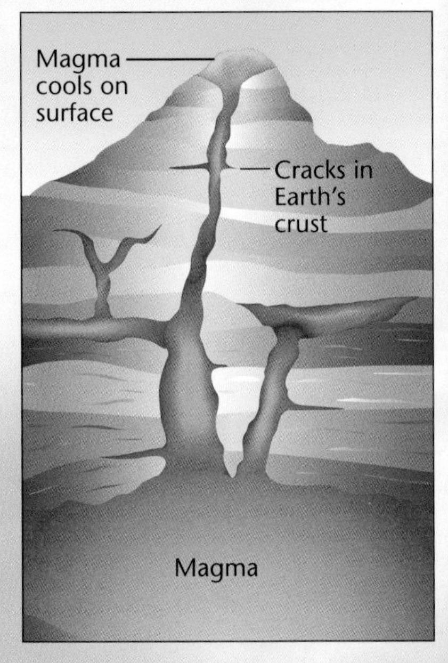

Molten magma is less dense than the surrounding crust. It rises up through cracks and cools to form minerals.

This picture shows huge clumps of shiny black obsidian jutting out of its surrounding rock.

MINERAL GROUPS

Minerals are divided into groups according to the elements which make them up. Minerals made of a single element are called **native elements**.

Pure silver on a piece of rock

This rock contains specks of pure gold.

Diamonds are crystals of pure carbon. Most are found in a rock called kimberlite, which forms under great heat and pressure.

Silicates, which contain **silica** (silicon combined with oxygen), are the largest group, making up 92% of minerals in the crust.

Beryl is a silicate made up of the elements silicon, oxygen, aluminum and beryllium.

Carbonates are minerals that contain elements combined with carbon and oxygen. They are the most abundant minerals after the silicates.

Smithsonite is zinc carbonate.

Malachite is copper carbonate. It is often polished and used in jewelry.

Halides are a group of minerals which contain halogen* elements.

Rock salt (halite) is formed when salt water evaporates.

Sulfides are a group of minerals that contain elements combined with sulfur.

Sphalerite is made of zinc and sulfur. Most of the world's zinc is mined from this mineral.

Phosphates are minerals formed when phosphorus reacts with oxygen and other elements.

Turquoise is a semi-precious mineral which is a phosphate of aluminum and copper.

Many elements combine with oxygen in the crust to form the group of minerals called **oxides**.

Hematite is a red iron oxide used to produce iron. It is also called "kidney-stone" because of its shape.

There are a number of other mineral groups containing oxygen. These all have names ending in "ate". The first part of their names (see below) show the other elements involved.

Mineral group	Element
Arsenates	Arsenic
Borates	Boron
Chromates	Chromium
Molybdates	Molybdenum
Nitrates	Nitrogen
Sulfates	Sulfur
Tungstates	Tungsten
Vanadates	Vanadium

See for yourself

Rocks are made up of a mixture of minerals. If you look at a rock with a magnifying glass, you can sometimes see the different minerals in it.

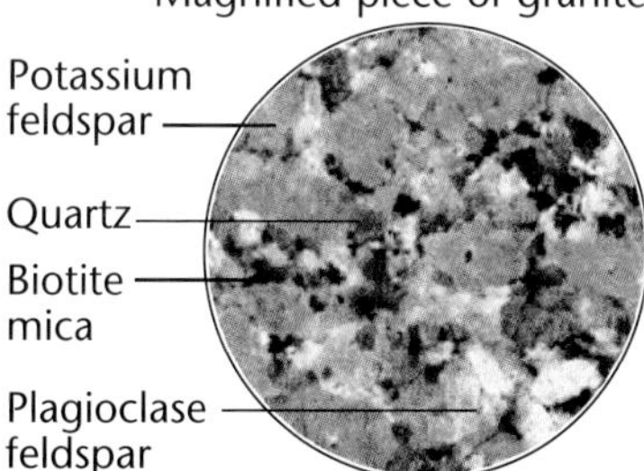

Magnified piece of granite

Internet links

- Go to **www.usborne-quicklinks.com** for a link to the **Infoplease Web site** for a definition of the term "mineral".
- Go to **www.usborne-quicklinks.com** for a link to the **Mineral and Gemstone Kingdom Web site** for a mineral guide.
- Go to **www.usborne-quicklinks.com** for a link to the **San Diego Natural History Museum Web site** for tips on growing and collecting minerals.
- Go to **www.usborne-quicklinks.com** for a link to the **Mineral Vug Web site** to see many different pictures in a mineral catalog.
- Go to **www.usborne-quicklinks.com** for a link to the **University of Texas Web site** for in-depth information on minerals.

* *Halogens, 46.*

THE PERIODIC TABLE

The **periodic table** is an arrangement of the elements placed in order of increasing atomic number (the number of protons in the nucleus). Each element is represented by a box containing its chemical symbol, atomic number and relative atomic mass (see far right). Some versions, such as the one shown here, also give the elements' names. New elements are added when they are discovered.

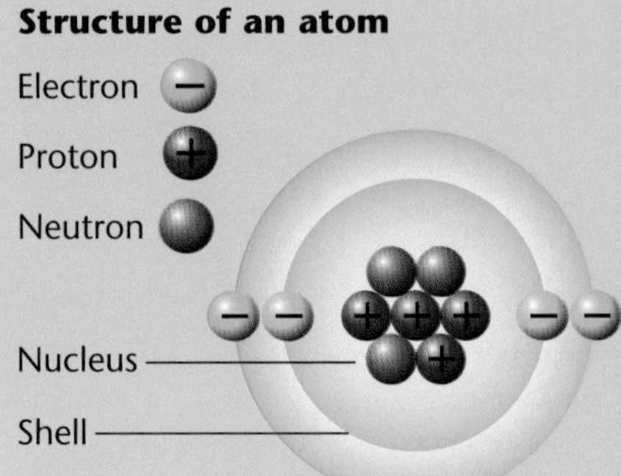

READING THE TABLE

The table is arranged into rows and columns. Looking at the table you will see that it has numbered rows (called **periods**) and columns (**groups**).

PERIODS

Each period is numbered, from 1–7. The atoms of all the elements in one period have the same number of shells, which contain electrons. For example, elements in period 2 have two shells and those in period 3 have three.

Moving from left to right across a period, each successive element has one more electron in the outer shell of its atoms. This leads to a fairly regular pattern of change in the chemical behavior of the elements across a period.

GROUPS

Each group has a Roman numeral, from I-VIII. Elements in the same group have the same number of electrons in their outer shell. This means that, chemically, they behave in similar ways.

Hydrogen is the lightest element. It has an atomic number of 1. It is not a metal so it is placed separately.

Key

Each element has a separate box in the periodic table containing the information below.

50	Atomic number
Sn	Chemical symbol
Tin	Name
118.7	Relative atomic mass

Period number — 1; Group number — I

Period number	I	II							
1	1 H Hydrogen 1.0								
2	3 Li Lithium 6.9	4 Be Beryllium 9.0							
3	11 Na Sodium 23.0	12 Mg Magnesium 24.3							
4	19 K Potassium 39.1	20 Ca Calcium 40.1	21 Sc Scandium 45.0	22 Ti Titanium 47.9	23 V Vanadium 50.9	24 Cr Chromium 52.0	25 Mn Manganese 54.9	26 Fe Iron 55.9	27 Co Cobalt 58.9
5	37 Rb Rubidium 85.5	38 Sr Strontium 87.6	39 Y Yttrium 88.9	40 Zr Zirconium 91.2	41 Nb Niobium 92.9	42 Mo Molybdenum 95.9	43 Tc Technetium (98)	44 Ru Ruthenium 101.1	45 Rh Rhodium 102.9
6	55 Cs Cesium 132.9	56 Ba Barium 137.3	71 Lu Lutetium 175.0	72 Hf Hafnium 178.5	73 Ta Tantalum 181.0	74 W Tungsten 183.8	75 Re Rhenium 186.2	76 Os Osmium 190.2	77 Ir Iridium 192.2
7	87 Fr Francium (223)	88 Ra Radium (226)	103 Lr Lawrencium (262)	104 Rf Rutherfordium (261)	105 Db Dubnium (262)	106 Sg Seaborgium (266)	107 Bh Bohrium (264)	108 Hs Hassium (269)	109 Mt Meitnerium (268)

The relative atomic masses for unstable, radioactive* elements are shown in brackets.

The elements with atomic numbers 57-70 belong to period 6.

The elements with atomic numbers 89-102 belong to period 7.

57 La Lanthanum 138.9	58 Ce Cerium 140.1	59 Pr Praseodymium 140.9	60 Nd Neodymium 144.2	61 Pm Promethium (145)	62 Sm Samarium 150.4	63 Eu Europium 152.0
89 Ac Actinium (227)	90 Th Thorium 232.0	91 Pa Protactinium 231.0	92 U Uranium 238.0	93 Np Neptunium (237)	94 Pu Plutonium (244)	95 Am Americium (243)

* *Radioactivity, 60.*

SIMILAR BEHAVIOR

On this periodic table, all elements that behave more-or-less in similar ways have the same colored background. The color-coding is explained here.

Non-metals
Mostly solid or gas, and non-shiny. Melt and boil at low temperatures.

Semi-metals
Also called metalloids, these have a mixture of the properties of metals and non-metals.

Metals
All are solid (except mercury, a liquid). Generally, they are shiny and have high melting points.

Transition metals are mostly hard and tough. Many are used in industry or jewelry.

Inner-transition metals are rare and tend to react easily with other elements, which makes them difficult to use in their natural state.

Transition metals

			III	IV	V	VI	VII	VIII
								2 He Helium 4
			5 B Boron 10.8	6 C Carbon 12.0	7 N Nitrogen 14.0	8 O Oxygen 16.0	9 F Fluorine 19.0	10 Ne Neon 20.2
			13 Al Aluminum 27.0	14 Si Silicon 28.1	15 P Phosphorus 31.0	16 S Sulfur 32.1	17 Cl Chlorine 35.5	18 Ar Argon 39.9
28 Ni Nickel 58.7	29 Cu Copper 63.5	30 Zn Zinc 65.4	31 Ga Gallium 69.7	32 Ge Germanium 72.6	33 As Arsenic 74.9	34 Se Selenium 79.0	35 Br Bromine 79.9	36 Kr Krypton 83.8
46 Pd Palladium 106.4	47 Ag Silver 107.9	48 Cd Cadmium 112.4	49 In Indium 114.8	50 Sn Tin 118.7	51 Sb Antimony 121.8	52 Te Tellurium 127.6	53 I Iodine 126.9	54 Xe Xenon 131.3
78 Pt Platinum 195.1	79 Au Gold 197.0	80 Hg Mercury 200.6	81 Tl Thallium 204.4	82 Pb Lead 207.2	83 Bi Bismuth 209.0	84 Po Polonium (209)	85 At Astatine (210)	86 Rn Radon (222)
110 Uun Ununnilium (269)	111 Uuu Unununium (272)	112 Uub Ununbium (277)		114 Uuq Ununquadium (285)		116 Uuh Ununhexium (289)		118 Uuo Ununoctium (293)

Inner-transition metals

64 Gd Gadolinium 157.2	65 Tb Terbium 158.9	66 Dy Dysprosium 162.5	67 Ho Holmium 164.9	68 Er Erbium 167.3	69 Tm Thulium 168.9	70 Yb Ytterbium 173.0
96 Cm Curium (247)	97 Bk Berkelium (247)	98 Cf Californium (251)	99 Es Einsteinium (252)	100 Fm Fermium (257)	101 Md Mendelevium (258)	102 No Nobelium (259)

Elements 57-70 are called the **lanthanides** or **rare earths**.

Elements 89-102 are called the **actinides** or **radioactive rare earths**.

RELATIVE ATOMIC MASS

Relative atomic mass is the average mass number of the atoms in a sample of an element. (The mass number is the total number of protons and neutrons in a nucleus.) Moving through the periodic table, elements are progressively heavier. For example, hydrogen (relative atomic mass: 1) is the lightest element. Ruthenium (101.1) is over a hundred times heavier.

GROUPS WITH NAMES

Some of the groups in the periodic table have names. For example, the metals in group I are all alkali metals and group II are alkaline earth metals. The elements in group VII are halogens and group VIII (sometimes called group 0) are called noble gases.

DIFFERENT VERSION

An alternative version of the periodic table shows it split into 18 groups rather than eight. This is achieved by treating each column in the transition metals section of the table as a separate group, numbered from 3-12. In this version, all groups are referred to by ordinary numbers, not Roman numerals.

Internet links

- Go to **www.usborne-quicklinks.com** for a link to the **WebElements Web site** to explore an interactive periodic table.
- Go to **www.usborne-quicklinks.com** for a link to the **Physics 2000 Web site** to find out about the periodic table's origins.
- Go to **www.usborne-quicklinks.com** for a link to the **FunBrain Web site** to play a periodic table game.

METALS

All the metal elements share certain properties. For example, they are shiny and they conduct electricity. They are classified according to the way they behave. For instance some, such as potassium and sodium, are very reactive and react violently with water and air, while others, such as gold, do not react at all.

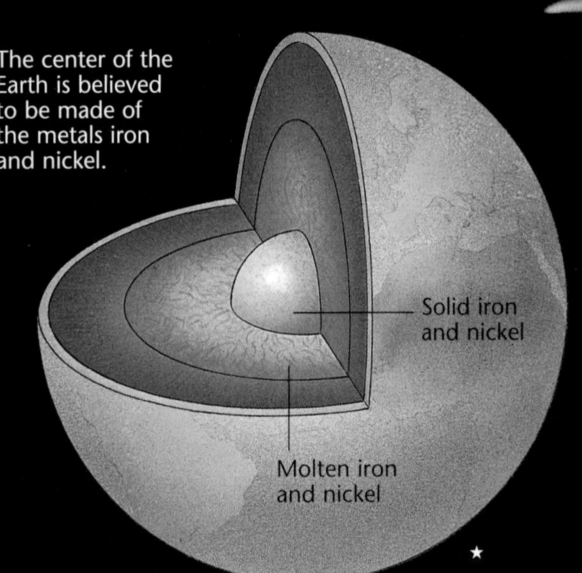

The center of the Earth is believed to be made of the metals iron and nickel.

Fireworks contain metal compounds that burn with brilliant colors.

PROPERTIES OF METALS

All metals, except for mercury, are solid at room temperature (20°C) and they are good conductors of electricity and heat. They are shiny when cut, and some, such as iron and nickel, are magnetic.

Metals that can be pulled out to make wire are described as **ductile**. Those that can be beaten flat are described as **malleable**.

Flat panel of malleable metal

Metal wire

THE REACTIVITY SERIES

The **reactivity series** is a list of metals showing how reactive they are. The position of each metal is decided by observing how they behave during reactions involving other metals. For instance, more reactive metals pull oxygen from less reactive metals.

Reactive metals are difficult to separate from the minerals in which they are found, while the least reactive metals can be found as pure metals.

Sodium and potassium are stored in oil as they react violently with air and water.

Copper is the least reactive metal that can be produced at a reasonable cost. It is used for pipes, hot-water tanks and electrical wiring.

Reactivity series

Most reactive
- Potassium
- Sodium
- Calcium
- Magnesium
- Aluminum
- Zinc
- Iron
- Tin
- Lead
- Copper
- Silver
- Gold
- Platinum

Least reactive

See for yourself

Metals conduct electricity, so a simple way to test whether something is made of metal is to see if you can pass an electric current through it. You can see this for yourself, using a simple electrical circuit.

You need:
3 pieces of insulated copper wire, each 8in long
6 volt lantern battery
3.5 volt bulb and holder

Using one wire, twist one end around a battery terminal and the other around one of the bulb holder's terminals. With a second wire, twist one end around the remaining terminal of the bulb holder. Twist one end of the third wire around the remaining battery terminal. (You can use tape to hold the wires in place.)

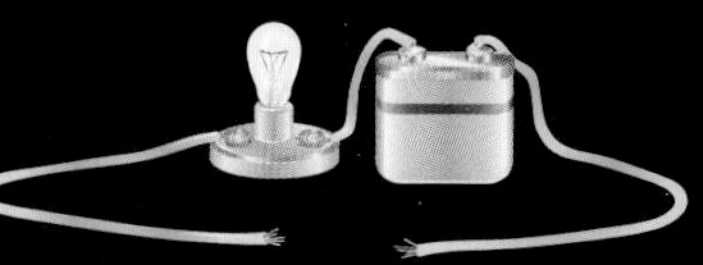

Carry the equipment around your home, touching the free wire ends to one object at a time. If the wires touch metal, electricity will flow through it and the light bulb will shine.

CAUTION
Never use an electrical outlet for this experiment. It is extremely dangerous.

FLAME TESTS

When some metals burn, they produce distinctive colored flames. Burning a substance can be used as a way to test for the presence of a particular metal. The substance is held in a flame on a piece of unreactive platinum wire.

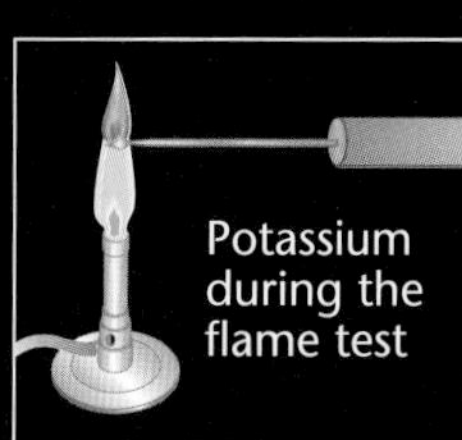

Potassium during the flame test

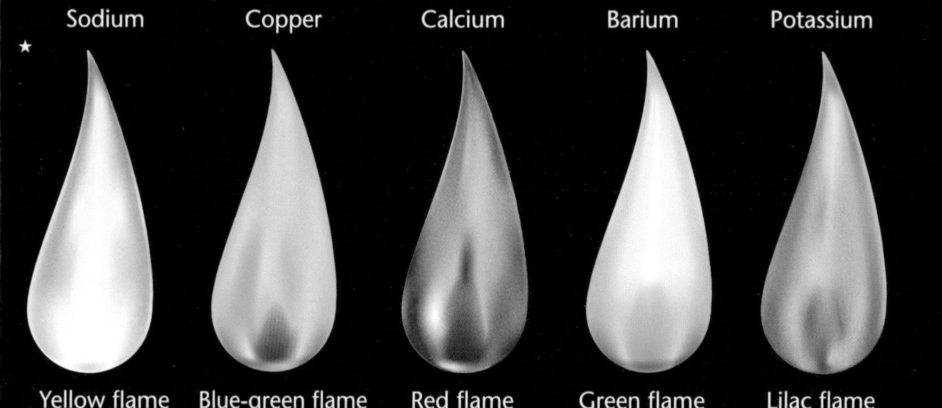

★

Internet links

• Go to **www.usborne-quicklinks.com** for a link to the **National Geographic Web site** to see how fireworks are made, then play the "Name that Boom" game.

• Go to **www.usborne-quicklinks.com** for a link to the **ScienceNet Web site** for useful information about metals and their properties.

• Go to **www.usborne-quicklinks.com** for a link to the **Fireworks Web site** to find out all about fireworks.

• Go to **www.usborne-quicklinks.com** for a link to the **Wonderful World of Minerals Web site** to learn basic facts about many common metals.

GROUPS OF METALS

Metals can be grouped according to their chemical properties and the way they behave. There are five main groups of metals, called noble metals, alkali metals, alkaline earth metals, poor metals and transition metals. Some of the noble metals are also transition metals.

The copper in this ore has not reacted with any other elements. Copper is a noble metal.

NOBLE METALS

Noble metals are those that can be found as pure metals, not as part of compounds, in the Earth's crust. These metals are copper, palladium, silver, platinum and gold.

The noble metals are all unreactive (see *The Reactivity Series*, page 28). They do not easily combine with other elements to form compounds.

Because they are unreactive, noble metals do not easily corrode and they are used for jewelry and coins. Gold is very unreactive and ancient gold objects are still shiny.

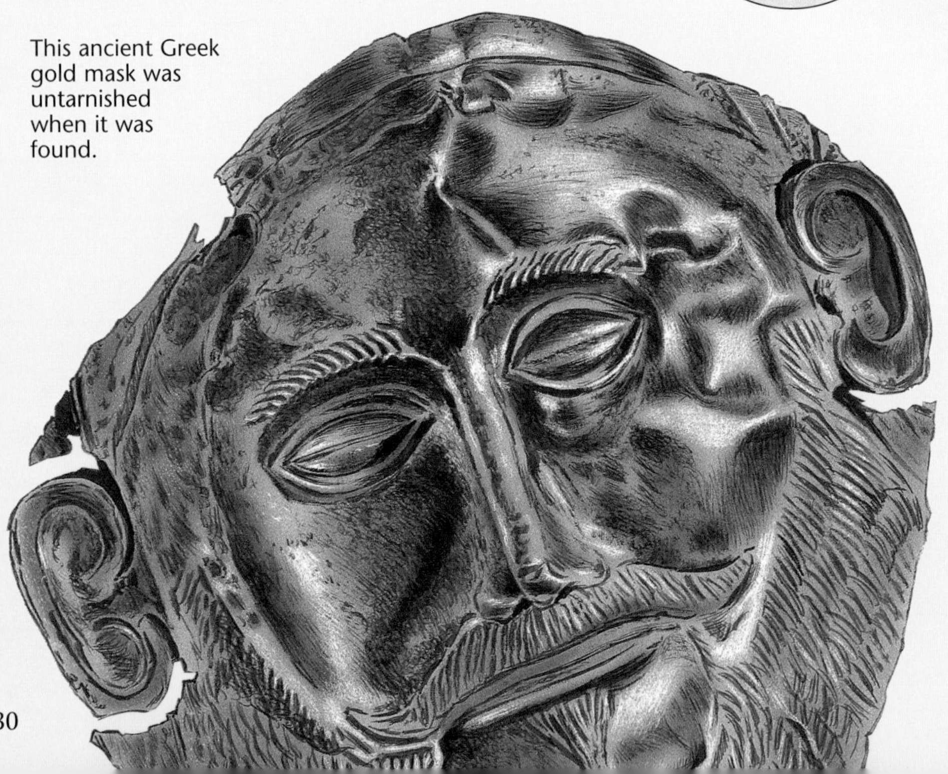

This ancient Greek gold mask was untarnished when it was found.

ALKALI METALS

The **alkali metals** are six very reactive metals, including sodium and potassium, that form group I of the periodic table. They have low melting points – potassium melts at 64°C – and they are soft and can be cut with a knife. They form alkaline* solutions when they react with water, which is why they are called alkali metals.

Potassium reacts violently with water, giving off hydrogen that bursts into lilac colored flames.

ALKALINE EARTH METALS

The **alkaline earth metals** are six metals, including magnesium, calcium and barium, that form group II of the periodic table. These metals are found in many different minerals in the Earth's crust. For example, calcium is found in calcite, which forms veins in limestone and chalk.

Alkaline earth metals are not as reactive as the alkali metals and they are harder and have higher melting points.

This shell contains large amounts of calcium, in the form of calcium carbonate.

Magnesium is found in chlorophyll, the green pigment needed by plants for photosynthesis.

* *Alkali, 58.*

TRANSITION METALS

The **transition metals** can be regarded as typical metals. They are strong, hard and shiny and have high melting points. They are less reactive than the alkali and alkaline earth metals.

Iron, gold, silver, chromium, nickel and copper are all transition metals. They are easy to shape and have many different industrial uses, both on their own and as alloys (see next page).

POOR METALS

The **poor metals** are a group of nine metals: aluminum, gallium, indium, tin, antimony, thallium, lead, bismuth and polonium. They are grouped to the right of the transition metals in the periodic table.

Poor metals are, in general, quite soft, and are not much use on their own. Many are used to make more useful substances, though.

Aluminum is one of the least dense metals. Lead, on the other hand, is very dense and is used in hospitals as a barrier against radiation and X-rays.

The frame of this bike is made mostly from titanium, a very light and extremely strong transition metal.

See for yourself

Many tooth fillings are made using mercury, a transition metal. A mercury-based filling (called **dental amalgam**) is inexpensive, hard wearing and easy for dentists to press into shape. Fillings made with mercury are a dull, light gray color.

You might also notice that some people have fillings or even an entire tooth made of a noble metal, particularly gold. Gold is used as a filling because it is harder wearing than a mercury filling, so it will last longer. Entire teeth are made of gold because they are practically unbreakable.

Internet links

- Go to **www.usborne-quicklinks.com** for a link to the **Reeko's Mad Scientist Lab Web site** to look at the color-coded groups of metals in the periodic table, then scroll down to find information about their properties.
- Go to **www.usborne-quicklinks.com** for a link to the **Metals Aerospace International Web site** for a dictionary of metal terminology.
- Go to **www.usborne-quicklinks.com** for a link to the **BBC Education Web site** to learn about Britain's metal-working heritage by clicking on "Midlands Metal Working Trail" and "North Wales Slate and Copper Trail"
- Go to **www.usborne-quicklinks.com** for a link to the **Copper Page Web site** for lots of information about copper.

ALLOYS

An **alloy** is any mixture of two or more metals, or a metal and another substance. Alloys are made because they combine the properties, such as lightness or strength, of the different metals which make them up.

Stainless steel, such as that used for knives and forks, is an alloy of steel, nickel and chromium.

This ship's propeller is made of bronze, an alloy of copper and tin. The bronze has been strengthened further by adding manganese.

Bronze is often used in ship building, because it is highly resistant to corrosion by sea water.

ADDING STRENGTH

In pure metals, the atoms are arranged tightly in rows. The rows can slide over each other and this makes the metal soft. Sudden pressure, however, can cause cracks to form across the rows, making the pure metal brittle.

Arrangement of atoms in pure metal

Slide

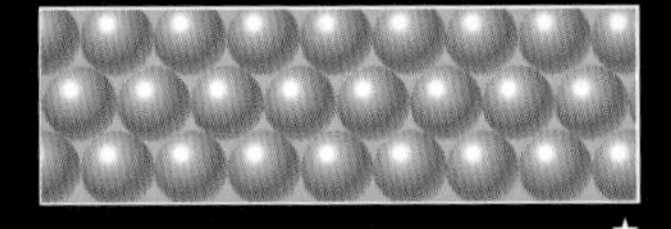

Slide

When another metal is added, its atoms help to strengthen the first metal. It does this by holding the parts of the metal together, so stopping its rows from sliding over each other.

Arrangement of atoms in an alloy

Atom of alloy

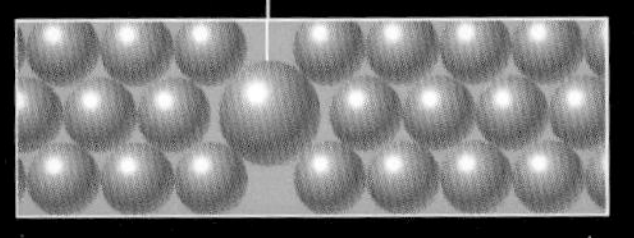

Atoms cannot slide.

ALLOY PROPERTIES

An alloy's properties depend on exactly what it is made of. Steel, for example, which is an alloy of iron and carbon, combines strength with ease of use. It can easily be worked into different shapes in a forge. It can also be melted without releasing poisonous fumes.

Steel's hard-wearing properties are increased by adding manganese. Steel-manganese alloys are used for industrial cutting equipment.

Railroad tracks are made of steel strengthened with manganese.

Some pure metals, such as gold and silver, are good at resisting corrosion, so they are ideal for use outside. But they are very expensive. Some alloys are just as good at resisting corrosion, yet are much cheaper to produce. Brass, an alloy of copper and zinc, is a good example.

Some alloys, such as bronze, a mixture of copper and tin, are easily shaped, even at room temperature. Because of this, bronze has been used for thousands of years to make decorative objects.

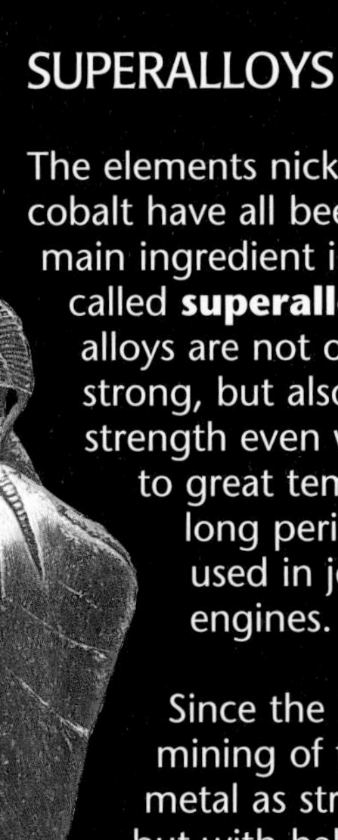

Ancient Greek bronze sculpture

STRONG, LIGHT ALLOYS

Like steel and brass, alloys of aluminum and magnesium, for example duralumin, are strong and corrosion resistant. But they are also much lighter. They are used for aircraft and bicycle frames.

Most modern jets are made from alloys of aluminum or super-strong titanium.

Metallurgists (scientists who study metals) have discovered that metals are often strongest if they are alloyed with only very tiny amounts of other substances. This has made it possible to create alloys that are very strong but still light.

This plane's engines are constructed from superalloys.

SUPERALLOYS

The elements nickel, iron and cobalt have all been used as the main ingredient in what are called **superalloys**. These alloys are not only extremely strong, but also retain their strength even when exposed to great temperatures for long periods. They are used in jet and rocket engines.

Since the 1950s, the mining of titanium, a metal as strong as steel but with half its weight, has become affordable. Titanium is widely used in alloys that form the bodies of planes.

See for yourself

Next time you come across these everyday things, notice the useful qualities of the alloys of which they are made.

- Knives, forks and spoons made from **stainless steel**. Unlike silver, it doesn't tarnish.
- Door handles made from **brass.** This is shiny and decorative when polished.
- Bike frame made from **aluminum alloy**. This is strong but much lighter than a bike with a steel frame.
- Metal tools such as hammers, screwdrivers and wrenches made from **toughened steel**. They are practically unbreakable because the steel contains added quantities of vanadium or chromium. If they were not toughened, the tools would splinter or shatter dangerously when used.

Internet links

- Go to **www.usborne-quicklinks.com** for a link to the **Ontario Mining Association Web site** to take a tour of metals and minerals around the home.
- Go to **www.usborne-quicklinks.com** for a link to the **British Stainless Steel Association Web site** for lots of information about stainless steel.
- Go to **www.usborne-quicklinks.com** for a link to the **The Copper Page Web site** to find out about the uses of copper and copper alloys.
- Go to **www.usborne-quicklinks.com** for a link to the **International Titanium Association Web site** to find in-depth information about titanium and its alloys by clicking on "Ti Information".
- Go to **www.usborne-quicklinks.com** for a link to the **Infoplease Web site** for a concise encyclopedia article about alloys.

IRON AND STEEL

Nearly all iron mined from the Earth is found as an **ore** (that is, combined with another substance). Most of it is made into steel, which is used to make many useful things, ranging from paper clips and tools to frames for giant buildings.

ELEMENT OR ALLOY?

Iron is an element which is extracted mostly from an ore called hematite, a compound of iron and oxygen. **Steel** is an alloy (that is, a mixture) of iron, carbon and traces of other metals.

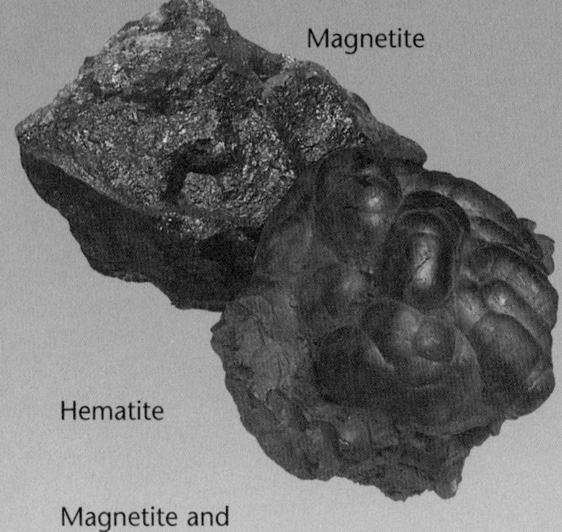

Magnetite and hematite are the two most common iron ores.

MAKING IRON

Iron is extracted from iron ore in a **blast furnace**. In the furnace, iron ore, limestone and **coke** (coal heated to burn off oils and leave carbon) are blasted with very hot air. This process is called **smelting**. The carbon combines with oxygen to form carbon monoxide. The carbon monoxide then becomes carbon dioxide by pulling the oxygen away from the iron ore. This is an example of a reduction* reaction.

The iron extracted from iron ore contains some left-over carbon (about 4%) from the smelting process, plus other impurities such as sulfur. Called **pig iron**, it is used to make cast iron, or refined further to make steel.

The smelting process

1. Iron ore, coke and limestone are fed into the blast furnace. The limestone reacts with the impurities in the ore to produce waste that is called **slag**.

2. Hot air is blasted into the furnace. It reacts with the carbon to form carbon monoxide. This reaction raises the temperature to about 3,600°F. Then the carbon monoxide reacts with the oxygen in the ore leaving the metal free.

3. Molten iron is tapped off here.

4. Molten slag runs out near the bottom of the furnace. This is used in road making.

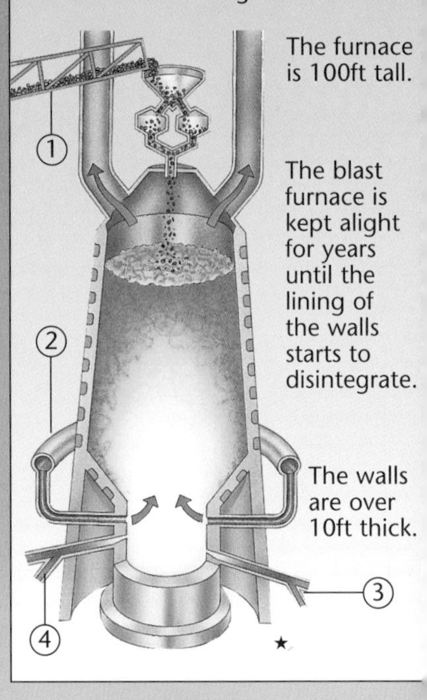

The furnace is 100ft tall.
The blast furnace is kept alight for years until the lining of the walls starts to disintegrate.
The walls are over 10ft thick.
1
2
3
4

The steel frame of this unfinished building will eventually be covered with concrete panels, to create a huge office block like those in the background.

* Reduction, 60.

MAKING STEEL

Steel is made of iron that has been through a blast furnace, with other elements added to make it stronger. To make steel, molten iron is blasted with oxygen, removing more carbon. The oxygen combines with carbon in the iron to form the gas carbon monoxide, which is collected and used as fuel. At the end, the steel may contain as little as 0.04% carbon, although different grades of steel contain different amounts.

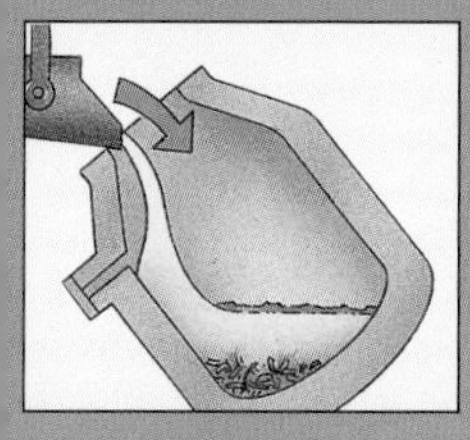

To convert iron to steel, molten iron is poured into a furnace called a **converter**.

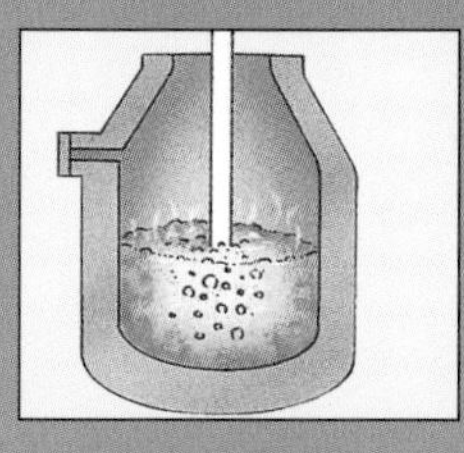

A high-pressure jet of almost pure oxygen is blasted into the converter. The oxygen combines with the carbon, forming carbon monoxide.

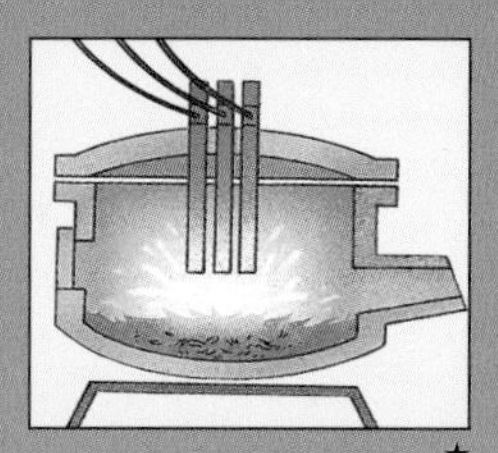

Steel is also made by melting down scrap steel in an **electric arc furnace**. The metal is melted by a powerful current of electricity.

See for yourself

Look out for iron and steel objects around your home. You can test to see if an object really is made of iron or steel (and not some other metal) by holding a magnet near it. If the object contains iron or steel, it will be attracted to the magnet.

Here are a few examples of things that you may have around your home, which you could test for their iron content.

- Metal door handles
- Hinges
- Knives and forks
- Garden gate
- Washing machine
- Bath
- Bike parts
- Food mixer
- Eyeglasses
- Belt buckles
- Faucets
- Radiators

CAUTION
Do not put a magnet near computers, television sets or watches. You could damage them.

The type of steel used for tools contains up to 1% carbon. This steel is very rigid but brittle. Chromium and vanadium are added to make the tools strong.

Steel that is used in construction, like the frame shown here, is painted to protect it from rusting before it is covered over by the rest of the building. A rusting frame would be dangerously weak.

Steel paper clips contain about 0.08% carbon. This makes them bendable. Those shown here are plastic coated to make them eye-catching.

Internet links

- Go to **www.usborne-quicklinks.com** for a link to the **BBC Online Web site** to see an animated blast furnace.
- Go to **www.usborne-quicklinks.com** for a link to the **How Stuff Works Web site** for information about iron and steel.
- Go to **www.usborne-quicklinks.com** for a link to the **ISSI Worldsteel Web site** for basic facts about stainless steel.
- Go to **www.usborne-quicklinks.com** for a link to the **CanCentral Web site** to learn about cans and the canning industry.
- Go to **www.usborne-quicklinks.com** for a link to the **British Metals Federation Web site** to discover how metals are recycled by clicking on "Tim Can".

MAIN METALS AND ALLOYS

There are 65 metals that naturally exist on Earth. Of these, just 20 are used, on their own or as part of an alloy, to produce nearly all manufactured, metal-based things. You can find out about those metals here, plus the five most common alloys, and see examples of how some of them are used.

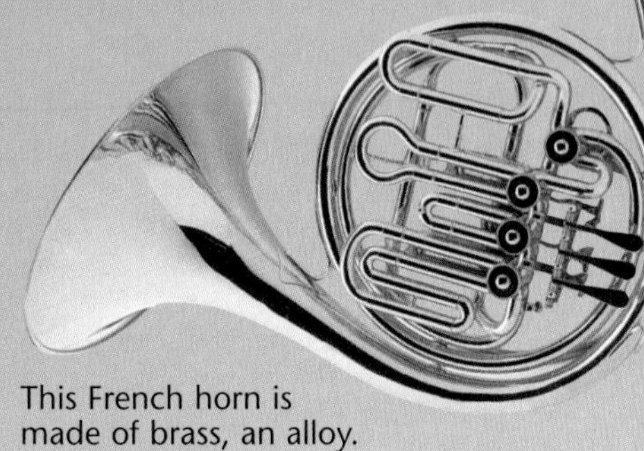

This French horn is made of brass, an alloy.

Aluminum
A very light, silvery-white metal that is resistant to corrosion. It is extracted from its ore, bauxite, by electrolysis*. Aluminum is used in overhead electric cables, aircraft, ships, cars, drink cans and kitchen foil.

Brass
An alloy of copper and zinc. It is easy to shape and is used for decorative ornaments, musical instruments, screws and tacks.

Bronze
An alloy of copper and tin known since ancient times. It resists corrosion and is easy to shape. Coins made of bronze are used as low-value currency in many countries.

Calcium
A malleable, silvery-white metal found in limestone and chalk. It also occurs in animal bones and teeth. It is used to make cement and high-grade steel.

Chromium
A hard, gray metal used to make stainless steel and for plating other metals to protect them or give them a shiny, reflective finish.

Copper
A malleable, reddish metal used to make electrical wires, hot water tanks and the alloys brass, bronze and cupronickel.

Cupronickel
An alloy made from copper and nickel from which most silver-colored coins are made.

Gold
A soft, unreactive, bright yellow element that is used for jewelry and in electronics.

Iron
A malleable, gray-white magnetic metal extracted mainly from the ore hematite by smelting in a blast furnace. It is used in building and engineering, and to make the alloy steel.

Lead
A heavy, malleable, poisonous blue-white metal extracted from the mineral galena and used in batteries, roofing and as a shield against radiation from X-rays.

Magnesium
A light, silvery-white metal that burns with a bright white flame. It is used in rescue flares and fireworks and in lightweight alloys.

Mercury
A heavy, silvery-white, poisonous liquid metal used in thermometers, dental amalgam for filling teeth, and in some explosives.

Platinum
A malleable, silvery-white unreactive metal used for making jewelry, in electronics and as a catalyst*.

Plutonium
A radioactive metal produced by bombarding uranium (see opposite) in nuclear reactors and used in nuclear weapons.

Potassium
A light, silvery, highly reactive metal. Potassium compounds are used in chemical fertilizers and to make glass.

Three million aluminum fasteners hold this jet's body together.

* Catalyst, Electrolysis, 58.

Silver
A malleable, gray-white metal that is a very good conductor of heat and electricity. It is used for making jewelry, silverware and photographic film.

Sodium
A very reactive, soft, silvery-white metal that occurs in common salt and is used in street lamps and in the chemical industry.

Solder
An alloy of tin and lead that has a low melting point and is used for joining wires in electronics.

Steel
An alloy of iron and carbon that is one of the most important materials in industry. Stainless steel, an alloy of steel and chromium, resists corrosion and is used in aerospace industries.

Tin
A soft, malleable, silvery-white metal. It is used for tin-plating steel to stop it from corroding, and in the alloys bronze, pewter and solder.

Titanium
A strong, white, malleable metal. It is very resistant to corrosion and is used in alloys for spacecraft, aircraft and bicycle frames.

Tungsten
A hard, gray-white metal. It is used for lamp filaments, in electronics, and in steel alloys for making sharp-edged cutting tools.

Uranium
A silvery-white, radioactive metal used as a source of nuclear energy and also in nuclear weapons.

Vanadium
A hard, white, poisonous metal used to increase the strength and hardness of steel alloys. A vanadium compound is used as a catalyst* for making sulfuric acid.

Zinc
A blue-white metal extracted from the mineral zinc blende (sphalerite). It is used as a coating on iron to prevent rusting (called galvanizing). It is also used in certain electric batteries and in alloys such as brass.

This Boeing 747 is built using a high-strength alloy that contains mostly aluminum – a very light metal. The jet engines are made of titanium, which is also light, but can easily withstand the enormous temperatures that are generated in the engines.

Internet links

- Go to **www.usborne-quicklinks.com** for a link to the **Minerals Council of Australia Web site** for information on the ten main minerals mined in Australia.
- Go to **www.usborne-quicklinks.com** for a link to the **ScienceNet Web site** to find out how we get metals from their ores.
- Go to **www.usborne-quicklinks.com** for a link to the **Gold Institute Web site** for fascinating facts about gold.
- Go to **www.usborne-quicklinks.com** for a link to the **Aluminium Federation Web site** to find out all about aluminum.

* *Catalyst, Galvanizing, 58.*

CORROSION

Corrosion is the chemical reaction that takes place when a metal is in contact with oxygen. The metal reacts with the oxygen to form a compound called an **oxide** on the surface of the metal. The metal becomes tarnished – that is, it loses its shine. Metals high in the reactivity series* corrode more quickly than less reactive metals.

Steel armor used to be rubbed with oil or beeswax to stop it from rusting.

USING METALS THAT CORRODE

Iron (from which steel is made) corrodes easily, but it is very strong and fairly easy to form into different shapes. It is ideal for building giant structures, such as bridges, but it has to be protected from corrosion, normally by painting it.

This bridge is protected from corrosion by painting it with phosphoric acid. The acid bonds to the metal and forms a protective coating, preventing rusting of the metal beneath. It is further protected by a layer of paint.

See for yourself

To remove the oxidized layer from a tarnished copper coin, leave it overnight in a glass containing a little vinegar. The acidic vinegar will react with the tarnish, removing it from the coin and exposing the copper alloy underneath. The coin will be left looking bright and shiny. Once it is back in the air though, it will corrode again, leaving a dull oxide layer on the surface.

* *Reactivity series, 28.*

EFFECTS OF CORROSION

When a metal corrodes, the surface becomes coated with a layer of oxide. On some, such as aluminum, this layer clings to the metal and protects it from further corrosion. On others this protective layer does not form. On iron and steel, for example, a flaky layer of **rust** (iron oxide) forms. This lifts away, allowing the metal beneath to corrode.

Aluminum immediately forms a layer of oxide on its surface. It is an ideal material for food trays, because it will not corrode further.

These steel drums were painted to protect them from rusting, but even a small scratch can let moisture under the paint, and rusting begins.

Moving parts, such as these gears, are coated with a layer of grease to stop them from rusting.

GALVANIZING

Galvanizing is a method of protecting steel by coating it with zinc. Zinc is more reactive than steel so oxygen reacts with it rather than the steel. Even if the layer of zinc is scratched, the oxygen in the air continues to react with the zinc rather than the steel.

Ships and oil rigs are protected by attaching a block of zinc or magnesium to them. This metal corrodes before the iron and is called the **sacrificial metal**.

Most modern cars are made from steel that has been galvanized. This stops them from rusting.

Internet links

- Go to **www.usborne-quicklinks.com** for a link to the **Copper Page Web site** to find out how copper has kept the Statue of Liberty beautiful for over 100 years.
- Go to **www.usborne-quicklinks.com** for a link to the **American Zinc Association Web site** where you can find detailed information about zinc, one of the main metals used to coat and protect other metals.
- Go to **www.usborne-quicklinks.com** for a link to the **How Stuff Works Web site** to learn what rust is and how it occurs.
- Go to **www.usborne-quicklinks.com** for a link to the **General Chemistry Online Web site** for the chemical reactions involved in rusting – plus their equations.
- Go to **www.usborne-quicklinks.com** for a link to the **Mad Scientist Experiments 2000 Web site** to see if metals rust more quickly in tap or salt water.

THE DISCOVERY OF METALS

People probably discovered how to extract metals from their ores by accident, when rocks containing a metal were heated with charcoal in fireplaces. A chemical reaction called reduction would have taken place which freed the metal from its ore. The same reaction is still used in blast furnaces (see page 34) to extract iron.

This decorated cauldron was made from bronze (a mixture of copper and tin) by the ancient Chinese, in around 1500BC.

THE FIRST METALS

The first metals worked by people were copper, gold and silver, probably because these are found as pure metals (see *Noble Metals*, page 30).

This golden cup was made in Northern Europe in about 3000BC.

Sumerian people in the Middle East made this golden dagger and sheath in about 4000BC.

Later, in about 3500BC, the Sumerians learned how to make bronze by combining copper and tin. Bronze is stronger than the pure metals.

Sumerian bronze bowl, made around 3000BC

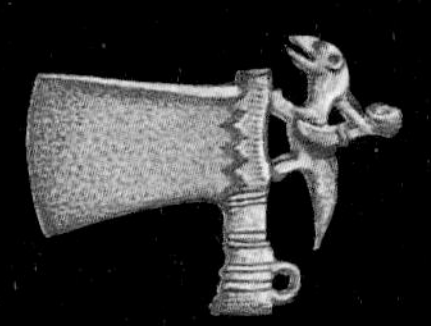

Bronze ax-head made in 500BC

Iron was not used until about 1350BC, probably because it needs much higher temperatures to separate it from its compounds.

Here, molten iron is being poured into a furnace which will produce steel.

NEW METALS

Until 1735, the only known metals were copper, silver, gold, iron, mercury, tin, zinc, bismuth, antimony and lead. Aluminum was discovered in 1825.

Nowadays scientists can create new metal elements, such as mendelevium, by bombarding atoms with electrons in a type of nuclear reactor called a **particle accelerator**. The atoms break apart under the bombardment, enabling scientists to get a glimpse of their structure.

This is part of a huge particle accelerator. It can be used to create new metals. These metals are unstable and break down in a very short time.

In this furnace, oxygen is blasted through the molten iron. The oxygen removes carbon from the iron, leaving stee This photograph was taken in 1958, but th steel-making process has changed little since then.

Internet links

- Go to **www.usborne-quicklinks.com** for a link to the **Carnegie Mellon University Web site** to read a short history of metals.
- Go to **www.usborne-quicklinks.com** for a link to the **Royal Mint Web site** to find out what percentage of different metals are found in British coins.
- Go to **www.usborne-quicklinks.com** for a link to the **Jewelry Supplier Web site** to find out about gold and copper, and their roles in history, culture and religion.
- Go to **www.usborne-quicklinks.com** for a link to the **Adopt an Element Web site** where you can find out about the discovery, properties and uses of iron.

RECYCLING METALS

Mining and extracting metals from ores is an expensive process. Fortunately, though, metals can be used again. The process of making them reusable is called **recycling**, and it is much cheaper than extracting metals from ores. Recycling is done by melting down used metal, to produce a metal that is almost as good as new. It can be done over and over again.

THE RECYCLING PROCESS

Before a metal can be recycled it needs to be collected and separated from any other types of metal. This ensures that the recycled metal is as pure as possible. It is then melted down and poured into molds. The metal cools into a solid block, ready to be made into a new, finished product.

WHICH METALS?

The most commonly recycled metals are steel and aluminum. However, other metals, such as copper, tin and lead and even precious metals, including gold, silver and platinum, are recycled too.

STEEL

Most steel for recycling comes from scrapped vehicles, such as cars and ships. Old industrial machinery is a good source too. Factories that use steel save offcuts and return them to steelworks to be melted and reused.

Some of the things that people throw out from their homes, such as old washing machines, also contain steel. Much of this can be recycled.

Here, red-hot recycled steel is being poured into a mold.

This giant electromagnet* is picking up vast quantities of junk iron and steel. The metal will be dumped in a blast furnace to be melted down.

* Electromagnet, 58.

ALUMINUM

Recycled aluminum accounts for about 30% of all aluminum in use. It takes only 5% as much energy to recycle aluminum as it does to extract it from its ore, **bauxite**.

The biggest source of aluminum for recycling is old drink cans. In North America, over 64 billion aluminum drink cans are recycled every year. More than half of the aluminum in a can has been recycled.

Crushed aluminum drink cans, ready to be melted down.

The materials being picked up by this giant magnet all contain iron.

See for yourself

Most cans have information printed on them stating what they are made of. Take a look the next time you have a can of drink. Most likely, the can will be made of aluminum, and it will also have a symbol on it, signifying that it is recyclable.

PRECIOUS METALS

Some industries use precious metals. The photographic industry, for example, uses lots of silver, which it recycles to cut costs and preserve resources. A great deal of recycled gold, silver and platinum comes from old jewelry and other ornaments.

When South America was invaded by Spaniards in the sixteenth century, thousands of golden treasures such as this were stolen, melted down and made into new ornaments.

Internet links

- Go to **www.usborne-quicklinks.com** for a link to the **New Steel Web site** for lots of information about steel and how it is recycled.
- Go to **www.usborne-quicklinks.com** for a link to the **Steel Recycling Institute Web site** where you can explore the process of recycling steel.
- Go to **www.usborne-quicklinks.com** for links to the **Alucan** and **Recycling in Springfield Illinois Web sites** where you can learn more about how aluminum is recycled.
- Go to **www.usborne-quicklinks.com** for a link to the **CanCentral Web site** to find out about the recycling of aluminum and steel cans in the USA.

HYDROGEN

Hydrogen is the lightest and most abundant element in the entire universe. The Sun and the stars are made of hydrogen gas, but on Earth, hydrogen is found only in compounds and does not occur naturally as a free element (that is, on its own).

Stars are globes of extremely hot hydrogen and other gases.

REACTIVE HYDROGEN

Hydrogen is very reactive. It burns easily and combines with many other elements. For example, water, the most plentiful compound on Earth, is made of hydrogen and oxygen. Fossil fuels, such as coal and oil, are compounds of hydrogen and carbon, and sugars and starch also contain hydrogen.

Sucrose ($C_{12}H_{22}O_{11}$) the sugar in candy, is a compound of carbon, hydrogen and oxygen.

See for yourself

If you pour yourself a glass of water, try to imagine what it is made of. Water (H_2O) is a compound of hydrogen (H) and oxygen (O). It contains twice as many hydrogen atoms as oxygen atoms. However, although there are more of them, the hydrogen atoms have such a small mass that they make up only 12.5% of the water's total mass.

The Sun is a massive ball of constantly exploding gases. It consists mostly of hydrogen and helium.

Occasionally, vast streams of burning hydrogen flare out from the Sun. These are **solar prominences**.

MAKING HYDROGEN

Hydrogen (H_2) can be made by reacting methane gas (CH_4) with steam (H_2O) as shown by the following chemical equation:

$$CH_4 + 2H_2O \rightarrow 4H_2 + CO_2$$

Most hydrogen made in this way is used to make ammonia (NH_3) for fertilizers. To make ammonia, hydrogen is combined with nitrogen using the **Haber process**, discovered by Fritz Haber in 1909.

THE HABER PROCESS

In the Haber process, nitrogen gas from the air and hydrogen extracted from methane (CH_4) are passed over a catalyst* of iron. Under very high pressure and at a high temperature, the gases react to produce ammonia gas (NH_3). This is cooled to form liquid ammonia.

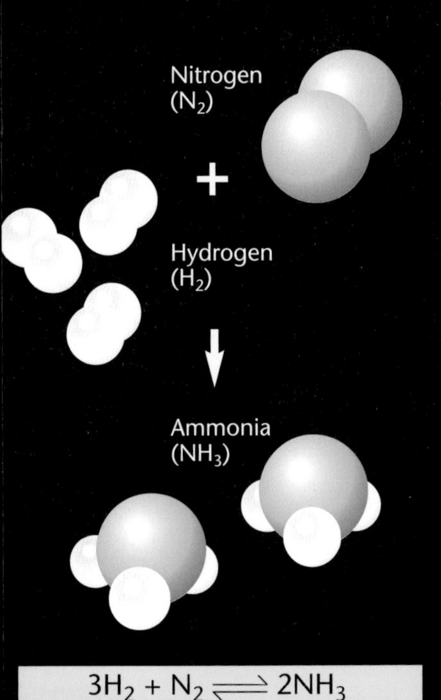

$$3H_2 + N_2 \rightleftharpoons 2NH_3$$

$\rightleftharpoons$ This symbol signifies that the reaction is reversible.

BURNING HYDROGEN

If hydrogen is mixed with air and then lit, it explodes. This can be used in the laboratory as a test for small amounts of gas. If the gas is hydrogen, it makes a little pop.

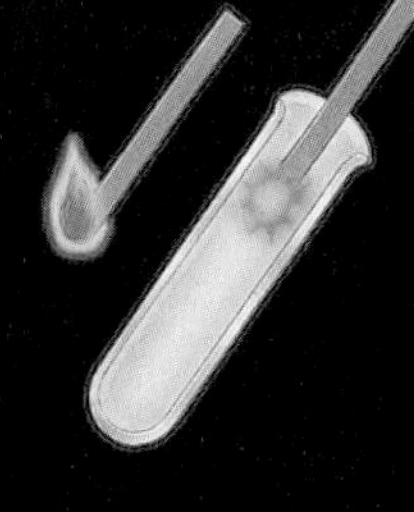

Hydrogen gas makes a small pop when tested with a burning splint.

If pure hydrogen (H_2) is burned in air or oxygen (O_2), it burns quietly with a blue flame and forms steam, as shown in this equation:

$$2H_2 + O_2 \rightarrow 2H_2O$$

In theory, hydrogen is an ideal fuel as it produces a lot of energy when it burns and the only product is water, which is not a pollutant. But at present it is not suitable as an everyday fuel because it is difficult to store and transport safely.

In 1937, the Hindenburg airship caught fire. It was filled with hydrogen, which exploded, killing 36 people.

ROCKET FUEL

Liquid hydrogen is used as a fuel for rockets. In order for the fuel to burn in space, where there is no oxygen, rockets also carry separate tanks of oxygen. The liquid hydrogen and oxygen are fed into a combustion chamber where they burn safely.

The fuel tanks have to be extremely strong to prevent the pressurized liquids from escaping.

Oxygen tank

Liquid hydrogen fuel tank

Internet links

- Go to **www.usborne-quicklinks.com** for a link to the **PBS Secrets of the Dead Web site** for a new theory about how the Hindenburg air disaster happened.
- Go to **www.usborne-quicklinks.com** for a link to the **National Hydrogen Association Web site** to find fascinating facts by clicking on "More About Hydrogen".
- Go to **www.usborne-quicklinks.com** for a link to the **U.S. Department of Energy Web site** for lots of information about hydrogen, including its production, handling and possible use as a fuel.
- Go to **www.usborne-quicklinks.com** for a link to the **New York Times Web site** to read about the development of hydrogen fuelled cars.
- Go to **www.usborne-quicklinks.com** for a link to the **Environmental Chemistry Web site** for detailed facts and figures about hydrogen – including its name in different languages.

* *Catalyst, 58.*

THE HALOGENS

The **halogens** are a group of five elements. They are fluorine, chlorine, bromine, iodine and astatine. They are all very reactive and poisonous and together form group 7 of the periodic table.

Halogen lamps contain compounds of bromine that make them shine really brightly.

FLUORINE

Fluorine is a poisonous gas. It is extracted from the mineral fluorite. **Fluorides** (non-poisonous compounds of fluorine) are added to toothpaste and drinking water to reduce tooth decay.

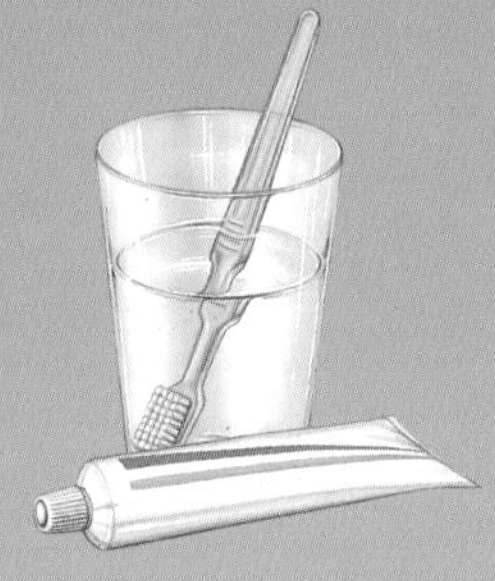

Toothpaste and water containing fluoride

Fluorine is also combined with carbon to make useful compounds called **fluorocarbons**. An example is PTFE (polytetrafluoroethene), which is used as a non-stick coating on frying pans and skis.

These skis have a coating of PTFE on their undersides. This non-stick layer helps them to slide freely over snow and ice.

CHLORINE

On its own, **chlorine** is a poisonous gas. It is very reactive and only occurs naturally in compounds such as sodium chloride (common salt). Chlorine is used as a disinfectant and to make hydrochloric acid and PVC (polyvinylchloride) plastic.

Compounds of chlorine have many uses. **Sodium hypochlorite**, for example, is used to make household bleach, and to bleach paper pulp so that it turns white.

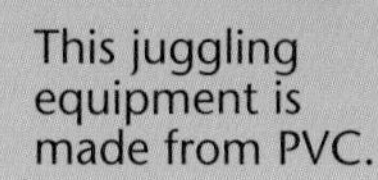

This juggling equipment is made from PVC.

Writing paper is bleached using sodium hypochlorite, a compound of chlorine.

This huge salt flat in South America contains sodium iodate. This is collected and used to produce the halogen iodine.

BROMINE

Bromine is a foul-smelling brown liquid. Traces of bromine are found in sea water and mineral springs. Compounds of bromine and one other element are called **bromides**. Silver bromide is used in photographic film.

When light hits silver bromide on photographic film, a reaction takes place in different layers of the film, creating various-colored patches.

Bromine compounds are used to make rat poisons and products that treat wood for termite infestation.

IODINE

Iodine is a purple-black solid. It is used in medicine, photography and dyes, and is produced in large quantities from sodium iodate.

Traces of iodine are found in foods, and without it the cells in our bodies would not be able to convert food into energy. However, large quantities of iodine are harmful.

Iodine is found in seaweed, and in vegetables and fruit.

ASTATINE

Astatine is an unstable, radioactive element. It is the heaviest of all halogens, but hardly any of it is found in nature. Scientists estimate that only about 1oz of astatine exists in the entire Earth's crust. They have been able to create more than 20 different astatine isotopes* during experiments, though.

See for yourself

You can buy iodine solution from a pharmacy and use it to test for the presence of starch. Drip some drops onto slices of food, such as raw potato, apple and a piece of bread. If starch is present, the food will turn blue-black very quickly.

This type of dropper is called a pipette.

Iodine tastes unpleasant, so make sure that you don't get any in your mouth.

Internet links

- Go to **www.usborne-quicklinks.com** for a link to the **Crest FamilyCare Center Web site** to find an activity showing how fluoride protects teeth (using any fluoride toothpaste).
- Go to **www.usborne-quicklinks.com** for a link to the **Chlorine Chemistry Council Web site** for lots of information about chlorine, including its many uses.
- Go to **www.usborne-quicklinks.com** for a link to the **How Stuff Works Web site** to discover the difference between halogen and normal light bulbs.
- Go to **www.usborne-quicklinks.com** for a link to the **WebElements Web site** where you can find detailed information about each halogen by clicking on its symbol (F, Cl, Br, I and At).

* *Isotopes, 11.*

CARBON

Carbon is a solid non-metallic element found in all living things. It occurs as a **free element** (on its own) mainly in the forms of hard, colorless diamond and crumbly black graphite.

FORMS OF CARBON

Carbon atoms can bond* together in different ways. These different forms are called **allotropes**. They contain the same types of atoms, but are bonded together in different ways. Carbon has three main allotropes – diamond, graphite and buckminsterfullerene.

Diamonds are cut in such a way that their surfaces split up light into the colors of the rainbow.

DIAMOND

In **diamond**, each carbon atom is bonded to four other atoms. This makes diamond very hard – it is the hardest substance found in nature. Diamond forms naturally as tetrahedral (four-sided) crystals.

Carbon atom in diamond

★

The crystal structure of a diamond sparkles brilliantly, and diamonds are valued for their beauty. They can be several different colors. The purest ones are transparent and are used to make jewelry.

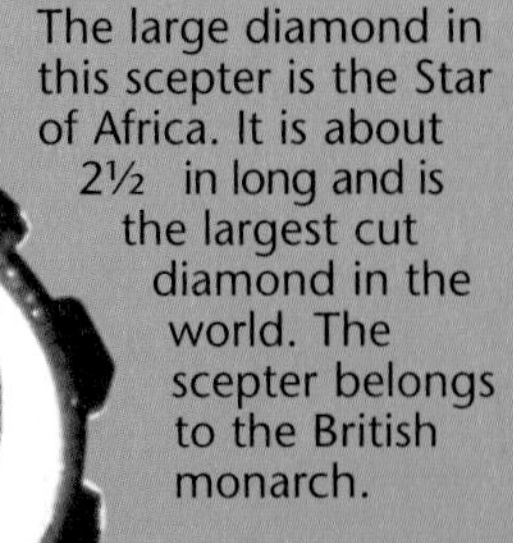

The large diamond in this scepter is the Star of Africa. It is about 2½ in long and is the largest cut diamond in the world. The scepter belongs to the British monarch.

* *Bond, 58.*

DIAMOND VARIETIES

Impure varieties of diamond, such as **carbonado** (also called **black diamond**) are valued in industry for their hardness. They are used in cutting and drilling equipment, as well as in some very accurate watches.

Naturally occuring varieties of diamonds are mined from the Earth, but diamonds can also be manufactured. These synthetic diamonds are created by mixing graphite with a catalyst* and subjecting it to great heat and pressure.

GRAPHITE

In **graphite** (sometimes called **plumbago**), each atom of carbon is bonded to three other atoms, arranged in a honeycomb-like network of plates that easily slide over each other. This makes graphite soft and flaky. The plate network is held together by weak forces.

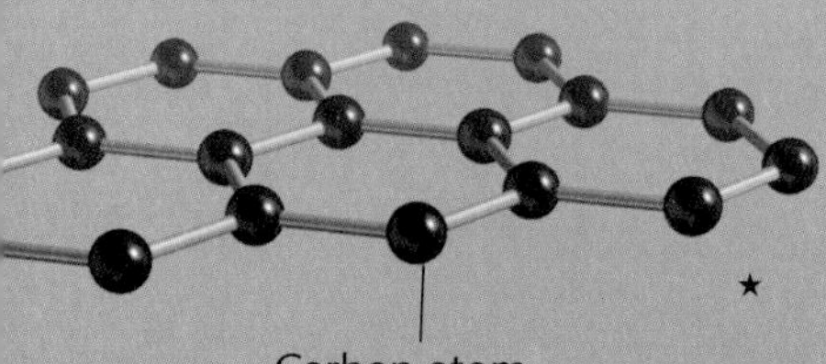

Carbon atom in graphite

The weak forces between the plates give graphite a very slippery structure. This makes graphite a very good lubricant, and it is used to reduce friction between the moving parts of machines. The weak forces also mean it is a good conductor of electricity so it is often used to make electrodes*.

Pencil "leads" are in fact made from powdered graphite mixed with clay. Soft pencils contain more graphite than hard ones.

BUCKMINSTERFULLERENE

Buckminsterfullerene is an allotrope of carbon discovered in 1985. Each molecule contains 60 carbon atoms linked in the shape of a hollow ball. It is formed by heating graphite in helium until it vaporizes, and then letting it cool and condense.

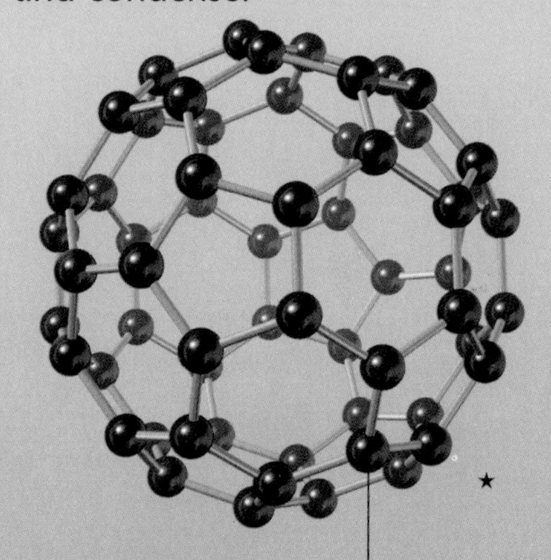

A buckminsterfullerene molecule

Buckminsterfullerene molecules are sometimes called **buckyballs**. Their atoms are arranged in a pattern of hexagons and pentagons similar to that on soccer balls.

Due to their robust spherical structure, buckyballs are really strong – a hundred times stronger than steel, but only a sixth of its weight.

Using a method similar to that for making buckyballs, scientists can also make tiny **nanotubes**. They hope to use them to build super-strong materials.

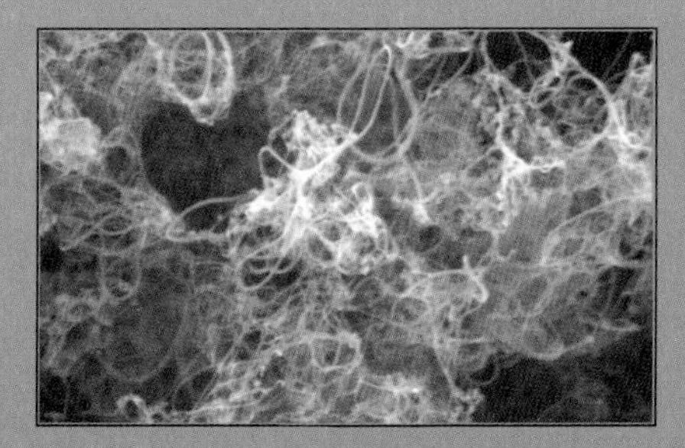

Nanotubes – made by vaporizing graphite with a laser and adding a metal catalyst*.

See for yourself

Take a look at your pencils, and compare the darkness of the lines that they make. A pencil containing more graphite than clay gives a darker, smudgier line than one that has more clay than graphite.

Letters and numbers printed on the side of a pencil indicate its graphite/clay content. A number followed by a B (for black) means that it has more graphite than clay. A number followed by an H (for hard) means it has more clay than graphite.

A medium pencil, one that gives lines that are neither dark nor light, has HB written on it.

This 4B pencil gives a soft, dark line. 9B is the darkest.

This 2H pencil gives a light, gray line. 9H is the lightest.

Internet links

- Go to **www.usborne-quicklinks.com** for a link to the **IN-VSEE Web site** to explore a World of Carbon.
- Go to **www.usborne-quicklinks.com** for a link to the **NASA Web site** to read all about "Buckyballs from Outer Space".
- Go to **www.usborne-quicklinks.com** for a link to the **Diamonds Web site** to find lots of information about diamonds.
- Go to **www.usborne-quicklinks.com** for a link to the **Diamond Deception Web site** for fascinating diamond facts.
- Go to **www.usborne-quicklinks.com** for a link to the **Beakman Web site** to build a diamond from marshmallows.
- Go to **www.usborne-quicklinks.com** for a link to the **Bristol University Web site** to learn about buckminsterfullerene.

* *Catalyst, Electrodes, 58.*

THE CARBON CYCLE

Most carbon atoms have existed since the world began. They circulate through animals, plants and the air in a process called the **carbon cycle**.

Plants use carbon dioxide to make carbon compounds by photosynthesis*. Animals eat plants (or other animals) and use the carbon compounds in their bodies. Carbon dioxide returns to the air when fuels burn and living things decay, and as a result of internal respiration, which is the way plants and animals break down sugars to release energy.

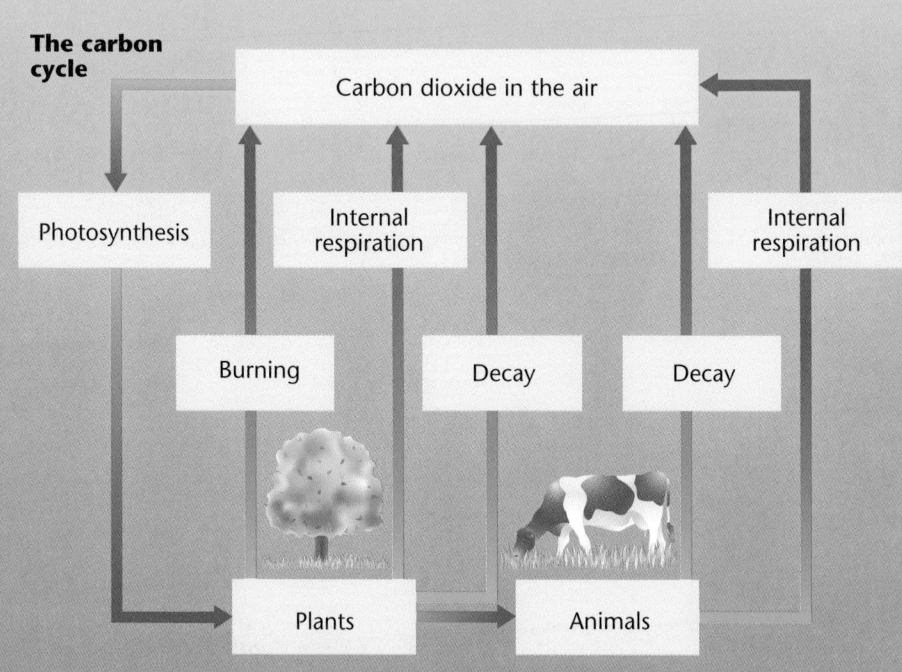

CARBON COMPOUNDS

Carbon atoms can bond with up to four other atoms, including other carbon atoms. This allows carbon to combine to form a vast number of different compounds. There are many more compounds of carbon than of any other element. All those compounds of carbon that are found in living things are called **organic compounds**.

Like all living things, both this kingfisher and the branch it is sitting on are made from compounds of carbon.

CARBON FIBERS

Silky threads of pure carbon, called **carbon fibers**, are used to reinforce plastics. This material is used to make lightweight boats and tennis racquets. A racing bike made of carbon fiber is eight times stronger than a steel one, but many times lighter.

This bike's frame is made from carbon fiber.

CARBON MIXTURES

Carbon can be found mixed with other elements and compounds. **Coal**, for example, is mainly carbon, but contains hydrogen, oxygen, nitrogen and sulfur. It is a **fossil fuel**, that is, a fuel formed over millions of years from the remains of plants.

There are three types of coal, containing varying amounts of carbon. **Lignite**, also called brown coal, only has 60-70% carbon. **Bituminous coal**, which is shiny and black, has more than 80%. **Anthracite** has more than 90% carbon.

Charcoal is another impure form of carbon. To make it, wood is heated in an airtight space. This removes the chemicals that produce wood smoke, leaving flaky black chunks of charcoal, which burn cleanly when ignited.

Unlike diamond and graphite, neither coal nor charcoal have a regular structure.

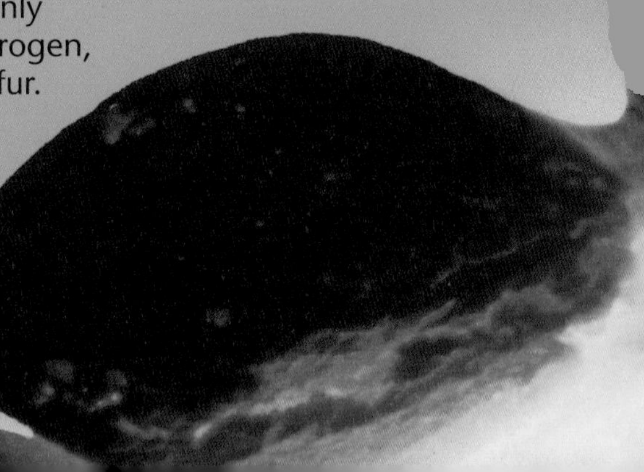

* Photosynthesis, 60.

USING COAL AND CHARCOAL

Coal is an important fuel. Over a third of the world's electricity is produced by power stations that burn coal. Lignite is cheap and plentiful, but produces a lot of pollution. Bituminous coal and anthracite are better since they cause less air pollution.

Power stations fuelled by coal can produce an average of 600 megawatts of electrical energy in an hour.

A form of charcoal called **activated charcoal** is used in filters and gas masks to remove poisonous fumes. It has countless tiny holes in its surface, which are ideal for trapping fumes. It is made by allowing charcoal to burn briefly in oxygen at the end of the charcoal-making process.

Charcoal is often used as a fuel on grills, and can be shaped into sticks to be used as an artists' drawing material.

See for yourself

The next time you see coal burning, try to imagine what is happening to the molecules that it is made of.

The heat gives the molecules enough energy to break apart. This gives off heat energy. As the bonds break, atoms, such as hydrogen, are freed from the molecules. These liberated atoms burn too, giving off additional heat.

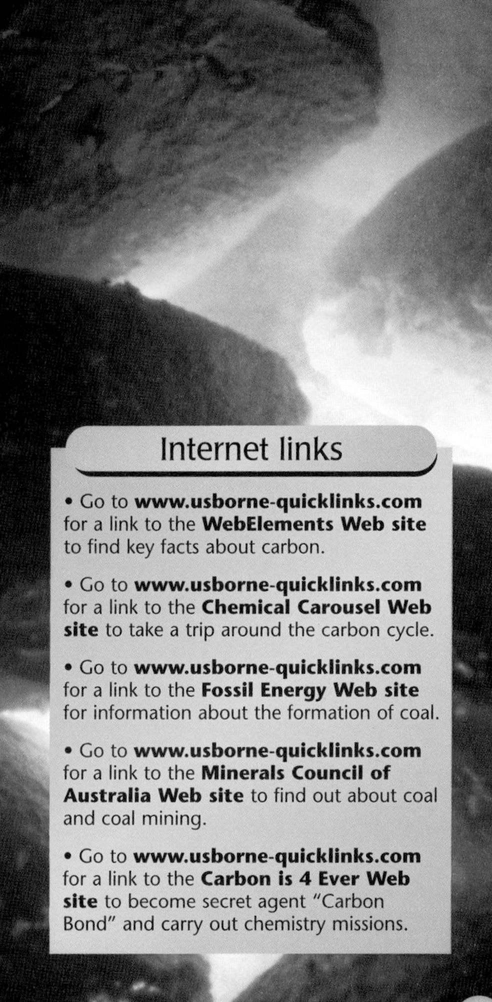

Charcoal burns without smoke. This makes it an ideal heat source for barbecue grills, because it cooks things without coating them in soot.

Internet links

- Go to **www.usborne-quicklinks.com** for a link to the **WebElements Web site** to find key facts about carbon.
- Go to **www.usborne-quicklinks.com** for a link to the **Chemical Carousel Web site** to take a trip around the carbon cycle.
- Go to **www.usborne-quicklinks.com** for a link to the **Fossil Energy Web site** for information about the formation of coal.
- Go to **www.usborne-quicklinks.com** for a link to the **Minerals Council of Australia Web site** to find out about coal and coal mining.
- Go to **www.usborne-quicklinks.com** for a link to the **Carbon is 4 Ever Web site** to become secret agent "Carbon Bond" and carry out chemistry missions.

SULFUR

The element **sulfur** is a bright yellow, crumbly solid. It is found in underground deposits in volcanic areas. It is also found in minerals such as iron pyrites and copper pyrites.

Pure sulfur

Iron pyrites, a compound of iron and sulfur

FORMS OF SULFUR

Sulfur molecules form in crooked rings of eight atoms, sometimes referred to as crowns. The rings can combine together in different ways to make two distinct crystal forms, known as allotropes.

Sulfur becomes a gas at 444°C. The molecules split apart and float freely, as shown here.

Most sulfur is found in the form of **rhombic sulfur**.

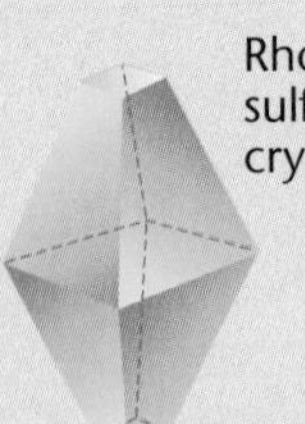

Rhombic sulfur crystal

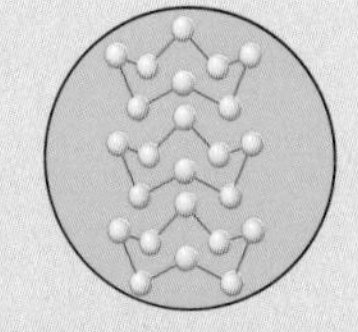

The molecules fit together closely in rhombic sulfur.

Above 96°C, **monoclinic sulfur** forms. Monoclinic sulfur crystals are long, thin and angular. They look a little like needles.

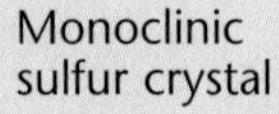

Monoclinic sulfur crystal

The molecules are less closely packed than in rhombic sulfur, so it is less dense.

PRODUCING SULFUR

Most sulfur is obtained from fossil fuels*. It is also extracted from underground deposits by melting it with pressurized steam. This is called the **Frasch process**.

USES OF SULFUR

One of the most important uses of sulfur is in the manufacture of **sulfuric acid**, which is used to make fertilizers, plastics and batteries. It is also used to **vulcanize** rubber (harden it), in black gunpowder and in medicines.

SULFUR DIOXIDE

Sulfur burns with a blue flame to form **sulfur dioxide**, a poisonous gas made of sulfur and oxygen. This gas is used to kill insects, as a fungicide and as a preservative for fruit.

Sulfur dioxide can be used to preserve the color of dried apricots.

Internet links

- Go to **www.usborne-quicklinks.com** for a link to the **ScienceNet Industrial Chemistry Web site** for a short article about sulfuric acid.
- Go to **www.usborne-quicklinks.com** for a link to the **Acid Rain - How's It Made? Web site**, where you can find out all about sulfur dioxide, one of the main causes of acid rain.
- Go to **www.usborne-quicklinks.com** for a link to the **Volcano World Web site** to explore a volcano crater and see sulfur crystals and gases from vents.

* *Fossil fuel, 58.*

PHOSPHORUS

Phosphorus is a non-metallic element. It occurs naturally in bones, teeth and the chemicals in the body that store energy. It is also found in the Earth, for example in the mineral apatite. Its most reactive form, white phosphorus, glows in the dark.

The minerals apatite (left) and turquoise (right) contain phosphorus.

FORMS OF PHOSPHORUS

Phosphorus occurs in three crystal forms, or allotropes.

White phosphorus is a poisonous, waxy, white solid that ignites easily when it is exposed to air.

Red phosphorus is a non-poisonous, dark red powder. It is made by heating white phosphorus without air. It is less reactive than white phosphorus.

Black phosphorus is made by heating white phosphorus under pressure using mercury as a catalyst*. Its name comes from its appearance, which is much like graphite. It is the least reactive form of phosphorus.

See for yourself

Take a look at the list of ingredients on a tube of toothpaste.

The list will probably include certain phosphates, such as sodium phosphate and trisodium phosphate. These are compounds that contain phosphorus.

These phosphates are used in toothpastes because they help to loosen stain-forming chemicals from your teeth, helping to keep them white.

USES OF PHOSPHORUS

One of the main uses of phosphorus is in the production of **phosphoric acid** (H_3PO_4). This is used to make iron and steel rust-proof, and in the making of carbonated drinks.

Phosphoric acid is used to add fizz and flavor to cola drinks.

Red phosphorus is used in matches, pesticides, alloys and distress flares.

White phosphorus is used in rat poison.

When a match is struck, the red phosphorus becomes white phosphorus, burning fiercely in the air.

Compounds of phosphorus and oxygen are called **phosphates**. Phosphates are important in animal and plant growth. They are added to animal feed and used to make fertilizers.

Farm crops, like this cabbage, are fed with large amounts of phosphate-rich fertilizers.

Internet links

- Go to **www.usborne-quicklinks.com** for a link to the **WebElements Web site** where you can find essential information about phosphorous.
- Go to **www.usborne-quicklinks.com** for a link to the **Amethyst Galleries Web site** where you can explore a searchable gallery of images and facts about phosphates.
- Go to **www.usborne-quicklinks.com** for a link to the **Arizona Dietetic Project Web site** where you can discover the part phosphorous plays in your diet.

* *Catalyst, 58.*

FACTS AND LISTS

PROPERTIES OF ELEMENTS

Here you will find data on the history and physical properties of most of the elements. Recently discovered elements such as ununhexium, about which little is known, are not included. "Year of discovery" is the year in which an element was first isolated. The figures shown are as accurate as possible, but there are different ways of measuring things, so you may find slightly different figures elsewhere.

Element	Melting point (°C)	Boiling point (°C)	Density (g/m^3)	Discovered by	Year of discovery
Actinium	1,050	3,200	10.1	Debierne	1899
Aluminium	660	2,467	2.7	Oersted	1825
Americium	994	2,607	11.7	Seaborg and others	1944
Antimony	630	1,750	6.62	-	Early historic
Argon	-189	-186	N/A	Rayleigh and Ramsay	1894
Arsenic	-	Sublimes	5.73	Albertus Magnus	1250
Barium	725	1,640	3.51	Davy	1808
Berkelium	986	Unknown	13.3	Seaborg and others	1949
Beryllium	1,278	2,970	1.85	Vauquelin	1798
Bismuth	271	560	9.78	Unknown	15th century
Boron	2,300	2,550	2.34	Gay-Lussac and Thénard; Davy	1808
Bromine	-7.2	59	3.12	Balard	1826
Cadmium	320	765	8.65	Stromeyer	1817
Caesium	28	678	1.9	Bunsen and Kirchoff	1860
Calcium	842	1,484	1.54	Davy	1808
Carbon*	3,730	4,830	3.51	-	Prehistoric
Cerium	798	3,257	6.78	Berzelius and Hisinger	1803
Chlorine	-101	-34	N/A	Scheele	1774
Chromium	1,857	2,672	7.19	Vauquelin	1797
Cobalt	1,495	2,870	8.7	Brandt	1735
Copper	1,083	2,567	8.89	-	Prehistoric
Curium	1,340	3,110	13.51	Seaborg and others	1944
Dysprosium	1,409	2,335	8.56	Boisbaudran	1886
Erbium	1,522	2,510	9.16	Mosander	1843
Europium	822	1,597	5.24	Demarcay	1901
Fluorine	-219	-188	N/A	Moissan	1886
Gadolinium	1,311	3,233	7.95	Marignac	1880
Gallium	29	2,403	5.93	Boisbaudran	1875
Germanium	937	2,830	5.4	Winkler	1886
Gold	1,064	2,807	19.3	-	Prehistoric
Hafnium	2,227	4,602	13.3	Coster and von Hevesy	1923
Helium	-272	-268	N/A	Ramsay	1895
Holmium	1,470	2,720	8.8	Delafontaine and Soret	1878
Hydrogen	-259	-252	N/A	Cavendish	1766
Indium	156	2,080	7.3	Reich and Richter	1863
Iodine	113	184	4.93	Cortois	1811
Iridium	2,410	4,130	22.4	Tennant	1803
Iron	1,535	2,750	7.85	-	Prehistoric
Krypton	-156	-152	N/A	Ramsay and Travers	1898
Lanthanum	920	3,454	6.19	Mosander	1839
Lead	327	1,740	11.3	-	Prehistoric
Lithium	180	1,347	0.53	Arfvedson	1817

** All entries refer to carbon in its graphite form.*

Element	Melting point (°C)	Boiling point (°C)	Density (g/m^3)	Discovered by	Year of discovery
Lutetium	1,656	3,315	9.84	Urbain	1907
Magnesium	648	1,090	1.74	Black	1755
Manganese	1,244	1,962	7.2	Gahn, Scheele and Bergman	1774
Mercury	-39	357	13.6	-	Prehistoric
Molybdenum	2,617	4,612	10.1	Scheele	1778
Neodymium	1,010	3,127	7.0	Von Welsbach	1885
Neon	-248	-246	N/A	Ramsay and Travers	1898
Neptunium	640	3,902	20.4	McMillan and Abelson	1940
Nickel	1,453	2,732	8.8	Cronstedt	1751
Niobium	2,468	4,742	8.57	Hatchett	1801
Nitrogen	-209	-196	N/A	Rutherford	1772
Osmium	3,045	5,027	22.5	Tennant	1803
Oxygen	-218	-183	N/A	Priestley	1774
Palladium	1,554	2,963	12.2	Wollaston	1803
Phosphorus	44	280	1.82	Brandt	1669
Platinum	1,772	3,827	21.5	de Ulloa	1735
Plutonium	641	3,232	19.8	Seaborg and others	1940
Polonium	254	962	9.4	Curie	1898
Potassium	63	759	0.86	Davy	1807
Praseodymium	931	3,212	6.78	Von Welsbach	1885
Promethium	1,080	2,460	7.3	Marinsky and others	1945
Radium	700	1,140	5	P. and M. Curie	1898
Radon	-71	-62	N/A	Dorn	1900
Rhenium	3,180	5,627	20.5	Noddack, Berg and Tacke	1925
Rhodium	1,966	3,727	12.4	Wollaston	1803
Rubidium	39	1,270	1.53	Bunsen and Kirchoff	1861
Ruthenium	2,310	3,900	12.3	Sniadecki	1808
Samarium	1,072	1,778	7.54	Boisbaudran	1879
Scandium	1,539	2,832	2.99	Nilson	1879
Selenium	217	685	4.79	Berzelius	1817
Silicon	1,410	2,355	2.35	Berzelius	1824
Silver	961	2,212	10.5	-	Prehistoric
Sodium	98	883	0.97	Davy	1807
Strontium	769	1,384	2.62	Davy	1808
Sulfur	119	444	2.07	-	Prehistoric
Tantalum	2,996	5,425	16.6	Ekeberg	1802
Technetium	2,172	4,877	11.5	Perrier and Segré	1937
Tellurium	449	990	6.2	Von Reichenstein	1782
Terbium	1,360	3,041	8.27	Mosander	1843
Thallium	303	1,457	11.8	Crookes	1861
Thorium	1,750	4,790	11.7	Berzelius	1828
Thulium	1,545	1,950	9.33	Cleve	1879
Tin	232	2,270	7.3	-	Prehistoric
Titanium	1,660	3,287	4.54	Gregor	1791
Tungsten	3,410	5,660	19.3	J. and F. Elhuijar	1783
Uranium	1,132	3,818	19.1	Peligot	1841
Vanadium	1,890	3,380	5.96	del Rio	1801
Xenon	-107	-111	N/A	Ramsay and Travers	1898
Ytterbium	824	1,193	6.98	Marignac	1878
Yttrium	1,523	3,337	4.34	Gadolin	1794
Zinc	419	907	7.1	-	Prehistoric
Zirconium	1,852	3,580	6.49	Klaproth	1789

LAWS

Avogadro's law All gases of the same volume at the same temperature and pressure must contain the same number of molecules.

Bernoulli's principle When the flow of a fluid (for example, air) gets faster, its pressure is reduced.

Boyle's law The pressure and volume of a gas at a constant temperature are inversely proportional.

Charles' law (or **law of volumes**) The volume of an ideal gas at a constant temperature is proportional to its kelvin temperature.

FAMOUS SCIENTISTS

Here are details of some of the scientists who have advanced our understanding of matter and the elements.

Avogadro, Amedeo (1776-1856) This Italian chemist devised Avogadro's law (see above).

Becquerel, Antoine (1852-1908) A French physicist who discovered radioactivity in 1896.

Bohr, Niels (1885-1962) A Danish physicist who applied the quantum theory of physics (see page 60) to Rutherford's structure of the atom in 1913.

Boyle, Robert (1627-1691) This Irish scientist proposed that matter is made up of tiny particles. He also formulated Boyle's law (see above).

Brown, Robert (1773-1858) A Scottish biologist who noted the apparently random motion of particles suspended in liquids. This is called Brownian motion.

Cavendish, Henry (1731-1810) This English chemist and physicist discovered hydrogen, the chemical make-up of air and water, and estimated the weight of the Earth.

Chadwick, James (1891-1974) An English physicist who worked on radioactivity and discovered the neutron.

Charles, Jacques (1746-1823) A French physicist who formulated Charles' law (see above).

Curie, Marie (1867-1934) A pioneering Polish scientist who carried out work on radiation and discovered the radioactive material radium in 1898.

Dalton, John (1766-1844) This English chemist suggested that elements are made of atoms which combine to form compounds.

Lavoisier, Antoine (1743-1794) This French lawyer and scientist named oxygen and hydrogen, and explained the role of oxygen in combustion.

Mendeleyev, Dmitri (1834-1907) This Russian chemist devised the periodic table of elements.

Rutherford, Ernest (1871-1937) A New Zealand-born physicist who demonstrated the structure of the atom.

MOHS HARDNESS SCALE

The hardness of minerals is measured on the **Mohs scale**, named after the German mineralogist, Friedrich Mohs (1773-1839). The scale has a sample mineral for each value, ranging from 1 – soft, crumbly talc, to 10 – diamond.

1. Talc
Very easily scratched with a fingernail.

2. Gypsum
Can be scratched with a fingernail.

3. Calcite
Very easily scratched with a knife, and just with a copper coin.

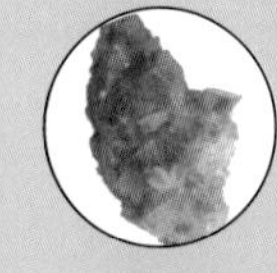

4. Fluorite
Easily scratched with a knife.

5. Apatite
Just scratched with a knife.

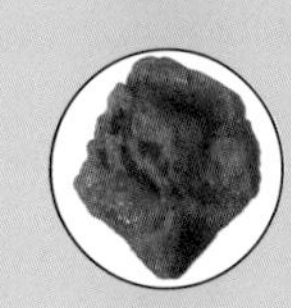

6. Orthoclase
Cannot be scratched with a knife. Just scratches glass.

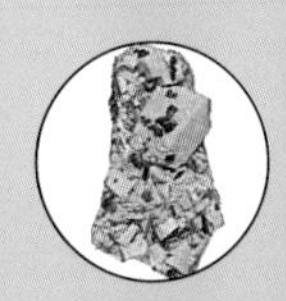

7. Quartz
Scratches glass easily.

8. Beryl or topaz
Scratches glass very easily.

9. Corundum
Cuts glass.

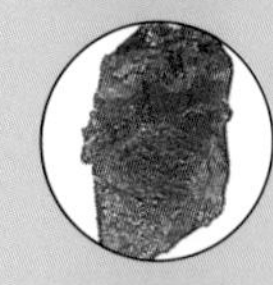

10. Diamond
Cuts glass very easily. Will scratch corundum.

TEST YOURSELF

1. Electrons are present:
A. only in liquid or solid matter
B. only in electrical conductors
C. in all forms of matter *(Page 8)*

2. An atom usually has equal numbers of:
A. neutrons and electrons
B. electrons and protons
C. protons and neutrons *(Page 9)*

3. The mass number of an atom of an element is the number of:
A. protons and neutrons
B. protons
C. electrons *(Page 10)*

4. Atoms which have the same number of protons and electrons but a different number of neutrons are:
A. isomers
B. isotopes
C. allotropes *(Page 11)*

5. The chemical symbol of iron is:
A. F
B. I
C. Fe *(Page 13)*

6. The chemical symbol of gold is:
A. Go
B. Au
C. Ag *(Page 13)*

7. The kinetic theory explains:
A. energy changes
B. moving objects
C. the properties of solids, liquids and gases *(Page 14)*

8. Sublimation occurs when:
A. solids change to gas
B. solids change to liquid
C. liquids change to gas *(Page 16)*

9. Condensation is when:
A. gas changes to liquid
B. gas changes to solid
C. liquid changes to gas *(Page 17)*

10. A substance is classified as a solid, liquid or gas depending on:
A. its state at 0°C
B. its state at 20°C
C. its state at 100°C *(Page 17)*

11. Gases have:
A. definite volume and shape
B. no definite volume and shape
C. definite volume but can change shape *(Page 20)*

12. How many elements are there?
A. about 20
B. about 50
C. about 100 *(Page 22)*

13. Almost all non-metals are:
A. liquids at room temperature
B. poor insulators
C. poor conductors *(Page 23)*

14. The most common element in the Earth's crust is:
A. aluminum
B. oxygen
C. silicon *(Page 24)*

15. The periodic table is organized into periods of elements which are:
A. arranged in columns
B. arranged in rows
C. arranged in clusters *(Page 26)*

16. The lightest element, with an atomic number of 1, is:
A. hydrogen
B. nitrogen
C. oxygen *(Page 26)*

17. Metals are ductile, which means they can be:
A. beaten flat into sheets
B. pulled out to make a wire
C. polished *(Page 28)*

18. In a flame test, potassium gives:
A. a red flame
B. an orange flame
C. a lilac flame *(Page 29)*

19. In water, alkali metals form:
A. acidic solutions
B. alkaline solutions
C. neutral solutions *(Page 30)*

20. Noble metals are:
A. always found in compounds
B. extremely reactive
C. very unreactive *(Page 30)*

21. Brass is a mixture of:
A. copper and zinc
B. copper and tin
C. copper and nickel *(Page 33)*

22. Bronze is a mixture of:
A. copper and zinc
B. copper and tin
C. copper and gold *(Page 33)*

23. Which gas is needed for corrosion to take place?
A. sulfur dioxide
B. carbon dioxide
C. oxygen *(Page 38)*

24. One of the oldest known metals is:
A. aluminum
B. gold
C. zinc *(Page 40)*

25. The most abundant element in the universe is:
A. aluminum
B. hydrogen
C. oxygen *(Page 44)*

26. Which element is not a halogen?
A. chlorine
B. iodine
C. phosphorus *(Pages 46-47, 53)*

27. Elements which exist in differently bonded forms are:
A. alloys
B. allotropes
C. isotopes *(Page 48)*

28. Which of the following substances is not a form of carbon?
A. diamond
B. sulfur
C. graphite *(Pages 48-49, 52)*

29. One of the main uses of sulfur is:
A. making sulfur dioxide
B. in food preservation
C. making sulfuric acid *(Page 52)*

30. Which of the following is not a form of phosphorus?
A. yellow
B. red
C. white *(Page 53)*

Answers
1.C 2.B 3.A 4.B 5.C 6.B 7.C 8.A 9.A 10.B 11.B 12.C 13.C 14.B 15.B 16.A 17.B 18.C 19.B 20.C 21.A 22.B 23.C 24.B 25.B 26.C 27.B 28.B 29.C 30.A

A-Z OF SCIENTIFIC TERMS

Acid A compound that contains hydrogen and dissolves in water to produce hydrogen ions.
actinides (also **radioactive rare earths** or **radioactive rare-earth metals**) A sub-group of the inner transition metals that includes actinium. All members of the sub-group have radioactive properties.
activated charcoal A form of charcoal (impure carbon) used to remove poisonous fumes, used, for example, in gas masks.
adhesion The attraction of the molecules of one substance to the molecules of another substance they are touching, rather than to each other (see also *cohesion*).
alkali Any **base** (the chemical opposite of an acid) that can dissolve in water, making an alkaline solution.
alkali metals The six very reactive metals that form group I of the periodic table. They are lithium, sodium, potassium, rubidium, cesium and francium.
alkaline earth metals The six metals that form group II of the periodic table. They are beryllium, magnesium, calcium, strontium, barium and radium.
allotropes Different forms in which certain elements, such as carbon, can exist. In each allotrope the (same) atoms are bonded together in a different way.
alloy A mixture of two or more metals, or a metal and a non-metal.
anodizing A method of coating a metal with a thin layer of its oxide using electrolysis, in order to prevent corrosion.
anthracite The purest form of coal, containing over 90% carbon.
atmosphere 1. The protective layer of air around the Earth that enables plants and animals to live. 2. The unit of atmospheric pressure at sea level.
atmospheric pressure The force of the air pressing down on the Earth's surface. At sea level, atmospheric pressure is one atmosphere (1 atm), also known as **standard pressure** and equivalent to 101,325 pascals.
atomic number The number of protons in the nucleus of an atom.
atoms The tiny particles from which elements are made. Each atom has a positively charged nucleus, consisting of protons and (except hydrogen) neutrons. This is usually balanced by enough negatively charged electrons to make the atom electrically neutral. See also *ion*; *isotopes*.

Ball-and-spoke model A way of representing molecules showing atoms as balls and the chemical bonds joining them as sticks.
bauxite The natural form of aluminum oxide. It is the ore from which aluminum is extracted.
black diamond See *carbonado*.
black phosphorus The least reactive allotrope of phosphorus, made by heating white phosphorus under pressure.
blast furnace A furnace used to smelt iron ore.
boiling point The temperature at which a substance turns from a liquid into a gas.
bond A force that holds together two or more atoms. Atoms make bonds in different ways, depending on their structure.
brown coal See *lignite*.
Brownian motion The movement of microscopic particles, such as dust, in liquids and gases, caused by constant collisions with rapidly moving molecules.
buckminsterfullerene A very strong allotrope of carbon with spherical molecules (often called buckyballs) containing 60 atoms.

Carbohydrates Organic compounds containing carbon, hydrogen and oxygen. They are broken down by all living things to give energy for life.
carbon An element found in compounds in all living things and also occurring naturally as diamond and graphite.
carbonado (or **black diamond**) An impure form of diamond used for cutting in industry because of its hardness.
carbon cycle The process by which carbon (as carbon dioxide) enters the food chain from the atmosphere through photosynthesis and returns to the atmosphere through respiration and decay. See also *internal respiration*.
carbon dioxide (CO_2) A gas with one carbon atom and two oxygen atoms in each molecule.
carbon fibers Silky threads of pure carbon, used to reinforce plastics.
catalyst A substance that changes the rate of a chemical reaction but is itself left unchanged.
catalytic converter A device fitted to cars which uses metal catalysts, for example platinum and rhodium, to remove toxic gases from the exhaust fumes.
CFCs (**chlorofluorocarbons**) Organic compounds of carbon, fluorine and chlorine, that are believed to damage the atmosphere.
chemical formula A combination of chemical symbols, showing the atoms of which a substance is made and their proportions.
chemical reaction An interaction between substances in which their atoms are rearranged to form new substances.
chemical symbol A shorthand way of representing a specific element in formulae and equations.
chlorofluorocarbons See *CFCs*.
cohesion The attraction of the molecules of a substance to each other, rather than to the molecules of a substance they are touching (see also *adhesion*).
combustion The scientific term for all forms of burning.
composites Synthetic materials, especially plastics, made up of different substances combined to improve their properties.
compound A substance made up of two or more elements whose atoms are chemically bonded.
condensation 1. The process of a gas cooling to form a liquid. 2. The droplets of liquid that form as a gas cools.
conductor A substance through which an electric current can flow, or through which heat can flow easily.
corrosion The process by which the surface of a metal reacts with oxygen to form the oxide of the metal, which then either protects the surface, or flakes off to cause damage. See also *rust*.
crust The outermost rock layer of the Earth. Below it is a layer called the **mantle**.
crystal A substance which has solidified in a definite geometrical form, with straight edges and flat surfaces.

Density A measure of the mass of a substance compared with its volume. See also *relative density*.
dental amalgam An alloy of mercury and copper used for filling tooth cavities.
diamond An allotrope of carbon with each atom linked to four others in a tight formation, forming extremely hard, four-sided crystals.
diffusion The spreading out of a gas to fill the space available.

ductile The term that describes a metal which can be stretched out to make wire.

Elastic The term describing something that can be stretched out of shape or size by a force, but returns to its original form when the force is removed (unless it exceeds its elastic limit). See also *plastic.*
elastic limit The point beyond which an elastic substance is altered, and weakened, permanently by stretching.
electric charge A property of matter which causes electric forces between particles. Opposite charges attract, while like charges repel each other.
electricity The effect caused by the presence or movement of electrically charged particles.
electrode In electrolysis, a conductor through which the current enters or leaves the solution.
electrolysis A method of splitting the elements in a compound by passing an electric current through it when it is molten or in a solution.
electromagnet A magnet which can be switched on and off by an electric current.
electron A negatively charged particle that moves around the nucleus of an atom.
electron cloud model A way of picturing electrons around an atom's nucleus. They are seen as moving randomly within cloud-like regions (also called **orbitals**).
electron configuration The number of electrons that exist in each of the shells around the nuclei of the atoms of a particular element.
electronics The use of electronic components to control the flow of a current around a circuit, making it do particular tasks.
electron shell A region (level) around an atom's nucleus in which a certain number of electrons can exist.
electroplating A method of covering an object with a thin layer of metal by electrolysis.
electrorefining A method of purifying metals using electrolysis.
element A substance made up of atoms which all have the same atomic number. It cannot be broken down by a chemical reaction to form simpler substances. See also *isotopes.*
Eureka can A container used to measure volumes of solid objects. It displaces the same volume of water as the volume of the solid put into it.
evaporation The process by which the surface molecules of a liquid escape into the air, becoming a vapor.

Ferromagnetic The term used to describe metals that can be magnetized easily.
flame test A test used to detect the presence of a particular metal by the color of its flame.
fossil fuel A fuel such as coal, oil or natural gas, that is formed from the fossilized remains of plants or animals.
Frasch process The process by which sulfur is extracted from underground, using pressurized steam.
freezing point The temperature at which a liquid cools sufficiently to become a solid (the same temperature as the melting point of the solid).

Galvanizing A method of protecting steel from rusting by coating it with zinc.
geysers Places where hot water and steam shoot out from below the surface of the Earth.
graphite (or **plumbago**) A soft, flaky allotrope of carbon in which each atom is linked to three others in a layered formation.
group In the periodic table, a column of elements whose atoms have the same number of electrons in their outer electron shell.

Haber process The method of producing ammonia by combining hydrogen and nitrogen.
halite See *rock salt.*
halogens The five poisonous and reactive non-metallic elements that make up group VII of the periodic table. They are fluorine, chlorine, bromine, iodine and astatine.
hard water Water which contains a lot of dissolved minerals from rocks it has flowed over. **Temporary hard water** contains minerals that can be removed by boiling. The minerals in **permanent hard water** cannot.
hydrometer A device used to measure the density of a liquid.

Inner transition metals A sub-group of the transition metals. Its members share similar properties, such as high reactivity.
insulator A substance that cannot conduct electric current, or does not conduct heat well.
integrated circuit See *silicon chip.*
internal respiration The process by which animals and plants use oxygen to break down their food, producing energy and releasing carbon dioxide.
ion An atom that has become electrically charged by gaining or losing one or more electrons. A negatively charged ion is called an **anion**, a positively charged ion is called a **cation**.

isotopes Different forms of the same element. The atoms of each isotope have the same number of protons, but a different number of neutrons, in their nuclei. So isotopes of an element have the same atomic number, but different mass numbers.

Kinetic theory The idea that all substances are made up of moving particles which have varying amounts of energy, giving rise to the different states of matter.

Lanthanides (also **rare-earths** or **rare-earth metals**) A sub-group of the inner transition metals that includes lanthanum.
lava Magma which has come out onto the surface of the Earth, for instance from a volcano.
lignite (or **brown coal**) The least pure form of coal, containing 60-70% carbon.

Magma Molten rock containing dissolved gases. Solid rocks containing minerals are formed when it wells up from below the Earth's surface, cools and solidifies. When it comes onto the surface, it is known as lava.
malleable The term used to describe metals that can be shaped by hammering.
mass The amount of matter contained in an object.
mass number The total number of protons and neutrons in the nucleus of an atom. The mass number of two atoms of the same element may be different, because they may be different isotopes.
mass spectrometer A device used to help identify atoms by sorting them by mass.
melting point The temperature at which a substance turns from a solid into a liquid (the same temperature as the freezing point of the liquid).
metalloids See *semi-metals.*
metallurgy The study of metals and their extraction.
metals Substances which make up three-quarters of all known elements. They are mostly shiny, solid at room temperature, have high melting points, and are good conductors of electricity and heat.
minerals The naturally occurring single and combined elements of which rocks are made.
mining The removal of minerals from the ground.
Mohs scale A scale by which the hardness of minerals is measured.
molecule The smallest particle of an element or compound that exists on its own and keeps its properties.

Nanotube A specially engineered microscopic fiber produced from graphite.
native elements Minerals consisting of a single element, e.g. gold.
neutron A subatomic particle with no electric charge. Neutrons form part of the nuclei of every atom (except those of hydrogen).
noble gases The six highly unreactive elements (all gases) present in the atmosphere which together form group VIII of the periodic table. They are helium, neon, argon, krypton, xenon and radon. The atoms of a seventh member of group VIII, ununoctium, can only be produced for fractions of seconds. It is not normally considered as a noble gas.
noble metal A metal, such as gold, that can be found naturally in a pure state.
non-metals Elements that are usually non-shiny, poor conductors, with low melting and boiling points. Eleven of the 16 known non-metals are gases at room temperature, four are solids and one (bromine) is a liquid.
nucleus The core section of an atom that contains protons and (except hydrogen) neutrons.

Orbital See *electron cloud model.*
ore A mineral compound from which a metal can be extracted.
organic compound A compound that contains the element carbon.
oxide A compound of oxygen and another element.
oxygen (O_2) A gas present in the Earth's atmosphere which is vital for life. It is also the most common element in the Earth's crust, found combined with other elements as oxides.

Particle accelerator A machine which accelerates charged subatomic particles to a great speed, in order that new particles produced by their collisions may be studied.
period A set of elements which have electrons in the same number of electron shells, shown as a row in the periodic table.
periodic table A systematic arrangement of the elements in order of increasing atomic number.
photon See *quantum theory.*
photosynthesis The process by which plants use energy from sunlight to power the conversion of water and carbon dioxide into carbohydrates.
physical states See *states of matter.*
plastic The term used to describe a substance which does not return to its original shape after stretching, but instead holds a new shape. See also *elastic.*
plumbago See *graphite.*
poor metals A group of nine metals, including aluminum and lead. They are mostly quite soft, and are usually combined with other substances to make alloys.
pressure The force exerted over a given area by a solid, liquid or gas.
proton A positively charged subatomic particle. Protons form part of the nucleus of every atom.

Quantum theory The idea that energy comes in tiny "packets", called quanta, which helps to explain the properties of electrons and the relationship of energy to matter. Quanta of electromagnetic radiation, such as light, are called photons.
quarks Particles which are believed to make up protons and neutrons.

Radioactive rare earths See *actinides.*
radioactivity The release of energy from the nuclei of unstable atoms. The energy, in the form of radiation, can be dangerous in large quantities.
radioisotope A radioactive isotope.
rare earths See *lanthanides.*
reactivity The tendency of a substance to react with other substances.
reactivity series A list of elements, usually metals, arranged in order of how easily they react with other substances.
recycling Making materials, such as metals, reusable by treating them in various ways.
red phosphorus A dark red, powdery allotrope of phosphorus, used to make matches.
reduction A chemical reaction in which a substance loses oxygen, or gains hydrogen or electrons. The opposite is **oxidation**, and the two processes normally occur together.
relative atomic mass The average mass number of the atoms in an exactly measured sample of an element (the mass numbers of the atoms may differ due to the presence of isotopes).
relative density A substance's density in relation to the density of water.
rhombic sulfur The most common allotrope of sulfur, with closely fitting molecules.
rock salt (or **halite**) The mineral form of common salt (sodium chloride).
rust Iron oxide that forms on the surface of iron, or an iron alloy, due to corrosion. It gradually flakes off, causing the iron to deteriorate.

Sacrificial metal A metal used to coat a less reactive metal, protecting it by corroding first.
semiconductor A type of material, usually a semi-metal, that acts as a conductor or as an insulator depending on its temperature.
semi-metals (or **metalloids**) Elemen which share some of the properties of metals and non-metals, and which can, depending on their temperature, be either conductors or insulators. They are all solids at room temperature. See also *semiconductor.*
shell See *electron shell.*
silicon chip (also **integrated circuit** or **chip**) A tiny piece of silicon with a complete electronic circuit etched onto it.
slag Impurities present in iron ore, left over as waste after smelting.
smelting The process of extracting a metal, usually iron, from its ore by heating to high temperatures, and reducing the ore.
solution A mixture that consists of a substance dissolved in a liquid.
space-filling model A way of representing molecules in which chemically bonded atoms are shown as balls clinging together.
stainless steel An alloy of steel and chromium that is both strong and resistant to corrosion.
standard pressure See *atmospheric pressure.*
states of matter (or **physical states**) The different forms in which a substance can exist. The three basic states are solid, liquid and gas.
subatomic particles Particles smaller than an atom, especially those of which atoms are made: protons, electrons and neutrons.
sublimation A change from solid to gas, or gas to solid, without going through liquid form.
superalloy An extremely strong, durable alloy that usually contains nickel, iron or cobalt.
surface tension The force of attraction that pulls together the molecules on the surface of a liquid.
synthetic The term describing compound materials that are made artificially by chemical reactions in a factory.

Transistor An electronic compone which acts as a switch by using a small current to control a larger o
transition metals The largest group metals, mostly hard and tough in nature with high melting points.

Volume The amount of space a substance occupies.
vulcanization A process used to strengthen rubber by heating it with sulfur.

Weight A measure of the strength of the pull of gravity on an object.
white phosphorus The most reactive allotrope of phosphorus, a poisonous white solid that ignites in air.

INDEX

You will find the main explanations of terms in the index on the pages shown in bold type. It may be useful to look at the other pages for further information.

Q

R

S

T

U

V

W

X

Y

Z

ACKNOWLEDGEMENTS

PHOTO CREDITS

(t = top, m = middle, b = bottom, l = left, r = right)

Corbis: **1** Eye Ubiquitous; **2-3** Science Pictures Ltd.; **10** (bl) Jim Sugar Photography; **12-13** (main) Digital Art; **14-15** W. Perry Conway; **16** (tl) Fukuhara Inc; **17** (m) Keren Su; **20-21** Roger Ressmeyer; **22** (bl) Jim Zuckerman; **24-25** Ric Ergenbright; **28-29** Buddy Mays; **30** (tr) Phil Schermeister; **32** Dan Guravich; **34-35** Dean Conger; **38-39** Roger Ressmeyer; **40-41** (main) Charles E. Rotkin; **41** (tr) Roger Ressmeyer; **43** (tr) James L. Amos; **46-47** (main) Charles O'Rear; **50-51** (main) Charles O'Rear; **53** (main) Owen Franken.
© **Digital Vision**: **4-5**; **6-7**;**17** (tm), (b); **19** (l); **20** (tr); **22-23**; **23** (m); **30** (br); **31**; **33** (ml); **35** (mr); **36** (tr); **36-37**; **39** (tr), (m), (mr); **42** (bm); **42-43**; **44** (t), (b); **46** (l); **50** (bl); **51** (tl); **54-55**.
Science Photo Library cover Sinclair Stammers; **NASA 49** (bm)
Crown copyright Historic Royal Palaces 48 (Reproduced by permission of Historic Royal Palaces under licence from the controller of Her Majesty's Stationery Office);
Lawson Mardon Star Ltd, Bridgnorth 22 (bm); **39** (l); **Robert M. Reed 25** (bm);
Ullstein Bilderdienst/Max Machon 45 (bm); **www.trekbikes.com 50** (m)

ILLUSTRATORS

Simone Abel, Sophie Allington, Rex Archer, Paul Bambrick, Jeremy Banks, Andrew Beckett, Joyce Bee, Stephen Bennett, Roland Berry, Gary Bines, Isabel Bowring, Trevor Boyer, John Brettoner, Peter Bull, Hilary Burn, Andy Burton, Terry Callcut, Kuo Kang Chen, Stephen Conlin, Sydney Cornfield, Dan Courtney, Steve Cross, Gordon Davies, Peter Dennis, Richard Draper, Brin Edwards, Sandra Fernandez, Denise Finney, John Francis, Mark Franklin, Nigel Frey, Giacinto Gaudenzi, Peter Geissler, Nick Gibbard, William Giles, David Goldston, Peter Goodwin, Jeremy Gower, Teri Gower, Terry Hadler, Bob Hersey, Nicholas Hewetson, Christine Howes, Inklink Firenze, Ian Jackson, Karen Johnson, Richard Johnson, Elaine Keenan, Aziz Khan, Stephen Kirk, Richard Lewington, Brian Lewis, Jason Lewis, Steve Lings, Rachel Lockwood, Kevin Lyles, Chris Lyon, Kevin Maddison, Janos Marffy, Andy Martin, Josephine Martin, Peter Massey, Rob McCaig, Joseph McEwan, David McGrail, Malcolm McGregor, Christina McInerney, Caroline McLean, Dee McLean, Annabel Milne, Sean Milne, Robert Morton, Louise Nevet, Martin Newton, Louise Nixon, Steve Page, Justine Peek, Mick Posen, Russell Punter, Barry Raynor, Mark Roberts, Andrew Robinson, Michael Roffe, Michelle Ross, Michael Saunders, John Scorey, John Shackell, Chris Shields, David Slinn, Guy Smith, Peter Stebbing, Robert Walster, Craig Warwick, Ross Watton, Phil Weare, Hans Wiborg-Jenssen, Sean Wilkinson, Ann Winterbottom, Gerald Wood, David Wright.

American editor: Carrie A. Seay